U0044974
惜墨如玉
TAIWAN
BLACK JADE
A UNIQUE NEW BREED OF GEMS
新 品 種 寶 石 ・ 臺 灣 墨 玉

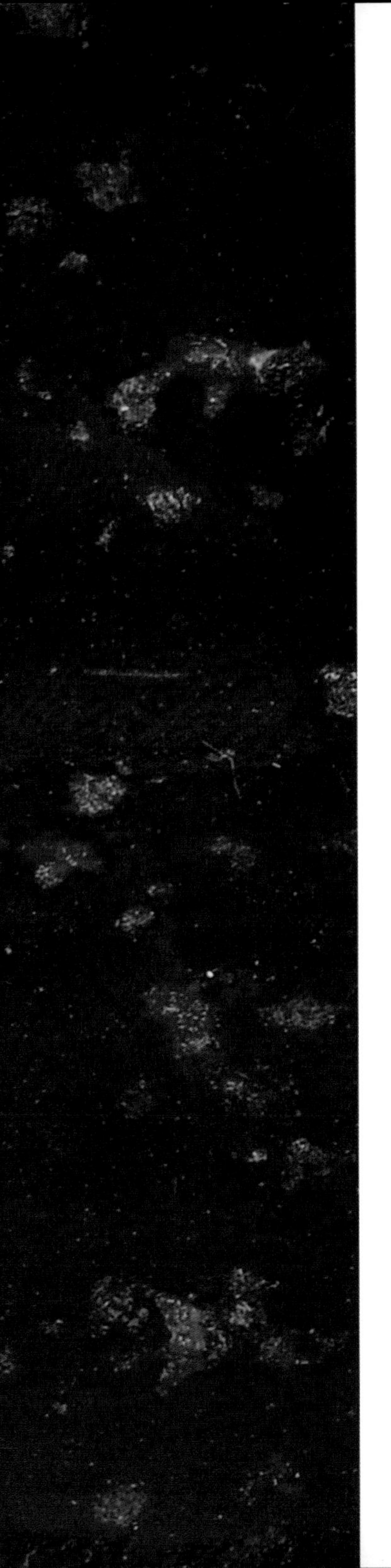

目錄

館長序 Preface 4
作者序 Preface 5
緣起 Prologue 6

第一章：玉文化與歷史 Jade Culture and History 7

玉文化與歷史 Jade Culture and History 8
史前時代的玉文化 Prehistoric Jade Culture 9
卑南文化出土玉石 Jadeware from the Peinan Site 10
歷史時代的玉文化 Jade Culture in the Historical Periods 11
文字中的玉 Jade-related Words in Chinese 13
何謂「玉」？ What is “Jade”? 14

第二章：臺灣玉 Taiwan Jade 15

世界聞名的臺灣玉 The World-renowned Taiwan Jade 16
臺灣玉的分類 Types of Taiwan Jade 17
臺灣玉的產狀與礦體分佈 Distribution of Taiwan Jade Deposits and Shapes of Ore Veins 18
臺灣玉的礦物特性 Mineral Properties of Taiwan Jade 19
最早發現的臺灣玉標本 The Earliest Sample of Taiwan Jade 20

第三章：臺灣墨玉 Taiwan Black Jade 21

臺灣墨玉的歷史溯源 History of Taiwan Black Jade 22
臺灣墨玉的命名 Naming of Taiwan Black Jade 23
臺灣墨玉的礦物特性 Mineral Properties of Taiwan Black Jade 24
臺灣墨玉的成因 Formation of Taiwan Black Jade 26
玉石萬花筒 Jade Kaleidoscope 27
世界重要玉石分布地 Major Deposits of Jade Stones around the World 28
臺灣玉與臺灣墨玉的化學成分 Taiwan Jade vs. Taiwan Black Jade: Chemical Composition 30

第四章：玉不琢不成器，玉石的加工 Processing of Jade 31

玉石的加工工序 Processing of Jade Stone 32

第五章：臺灣玉石精品 Taiwan Jade Boutique 33

傳統巧雕 Exquisite Jade Carving 34
宗教神學 Theological and metaphysical significances of Jade 40
時尚新設計 Innovative New Design 43
古文化今文創 Ancient Culture Modern Cultural Creativity 54

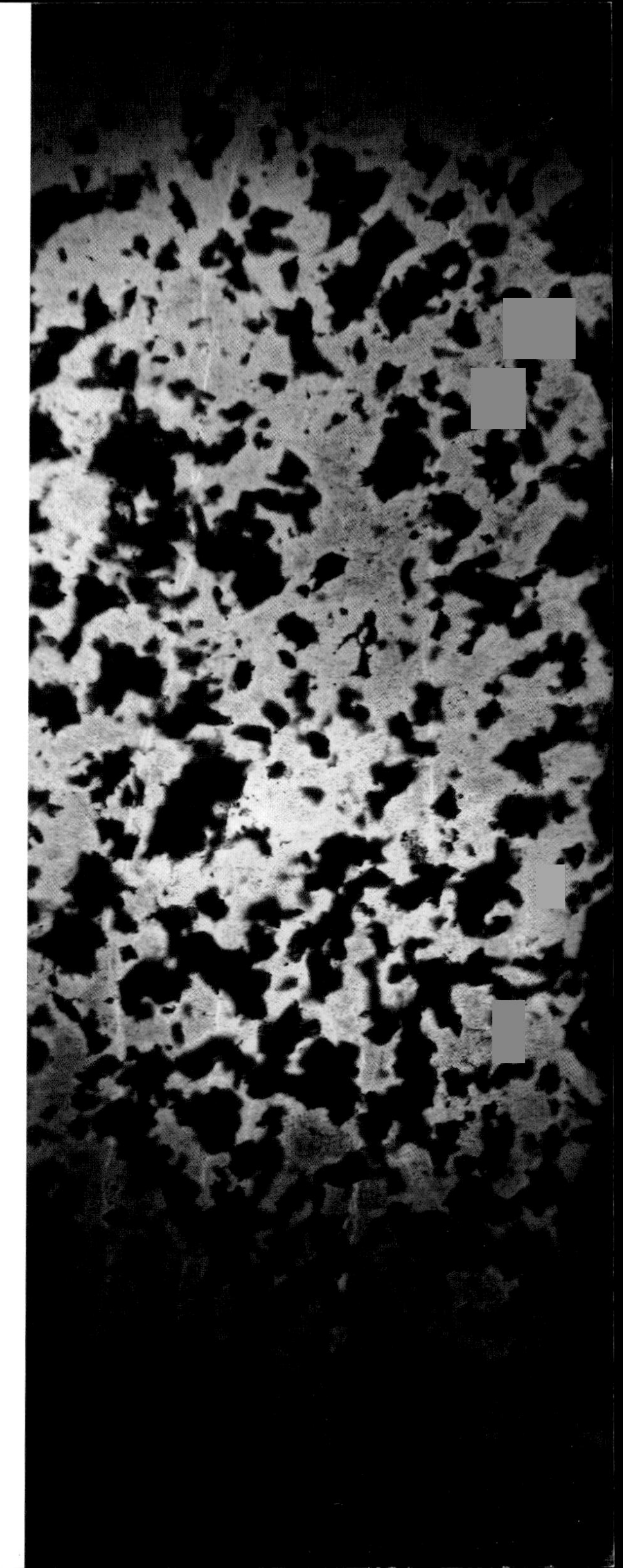

館長序 Preface

國立臺灣博物館是我國歷史最悠久的自然史博物館，對於自然史領域的地球科學中，有關寶石礦物的學術研究與推廣教育，一向不遺餘力。

本館繼 2012 年辦理「獨具慧『眼』－臺灣文石」的特展，帶領民眾認識澎湖和三峽地區所產令人驚艷的文石礦物之後，又於 2014 年積極與經濟部礦務局、國立東華大學藝術創意產業學系、國立臺灣史前文化博物館合辦，並得到國立臺灣大學地質科學系、經濟部中央地質調查所、中華民國鑛業協進會、天星礦場、臺灣玉發展協會等眾多研究單位與民間機構的齊力協助，再次推出以臺灣花蓮地區所產珍貴美麗且獨特的寶石－「臺灣墨玉」為主題的展覽。

臺灣素有「寶島」美稱，事實上臺灣島也確實蘊藏豐富多樣的寶石，包括臺灣玉、文石、玉髓、玫瑰石、紅珊瑚等等，其中臺灣玉是最具世界知名度的寶石，而且主要出產自臺灣花蓮的豐田地帶，這些綠色閃玉數量，曾多年名列為世界第一，所以「臺灣玉」幾乎可以說成是綠色閃玉的代名詞。

「臺灣墨玉」與臺灣玉共生，是另一種色澤獨具的玉石，亦是本館同仁在 2010 年間自行研究計畫中，利用館內的精密檢測儀器，經過詳實精確的礦物學研究後，所發現並命名的臺灣產新品種寶石，這在地球科學學術研究的領域，是一項十分重要的研究成果；為彰顯並推廣這項珍貴且重要的科學發現，本館 2014 年特別策劃大型展覽，以「惜墨如玉：新品種寶石－臺灣墨玉」為名，首次展出「臺灣墨玉」寶石的獨特樣貌，並搭配名聞遐邇精美的臺灣玉展品同時展出。

為配合「惜墨如玉：新品種寶石－臺灣墨玉」特展，本館特別發行專輯，此次因讀者熱烈的閱讀需求發行初版第三刷。本專輯內容由玉石的文化與歷史談起，介紹常見玉石的礦物學分類，最後深入了解到臺灣玉和臺灣墨玉的寶石特性，希望引領大家進一步認識這種極具發展潛力的臺灣本土玉石新品種－「臺灣墨玉」！

國立臺灣博物館館長

洪世佑 謹誌

作者序 Preface

臺灣，這個從十六世紀以來西方人眼中的「福爾摩沙」—美麗的寶島，蘊藏並生產許多寶石，比較常見的有閃玉(臺灣玉)、文石、玉髓(藍玉髓和紫玉髓)、珊瑚、玫瑰石、石榴子石和碧玉等，大多屬於半寶石類。其中臺灣玉與紅珊瑚兩項的產量，曾經在世界上佔有一席之地！以臺灣島的面積，相對於上述寶石的種類數量及某些寶石的產量，臺灣還真是一個「寶島」呢！

臺灣花蓮出產的綠色閃玉世界聞名，所以「臺灣玉」幾乎成了綠色閃玉的代名詞。除了閃玉之外，近來經濟部礦務局積極輔導玉石業者，開發出一種新品種的黑色蛇紋石玉，在強光下以黃、綠、黃綠色間夾黑色斑點呈現，因其具有獨特的岩理有別於中國的「岫岩玉」及「岫玉」，而且玉石蘊藏量相當驚人，極具開發的潛能。此種玉石經由臺博館研究人員於 2010 年進行礦物學研究發現：在野外地層中，雖然與知名的臺灣玉生長在一起，但卻是由不同礦物種類組成，因此化學成分也有差異。如果蛇紋岩中原來不透明的蛇紋石類礦物，經過適當的地質作用後，呈現透明或半透明，加上不透明的磁鐵礦，就形成了具有特殊美感的墨綠色半透明玉石，因此將它命名為「臺灣墨玉」，以便與世界其他寶石有所區分。

為了讓國人進一步了解此種臺灣本土所產的新品種玉石，國立臺灣博物館特別主辦「惜墨如玉:新品種寶石－臺灣墨玉」特展，介紹玉的文化與歷史、玉石的分類、臺灣玉及臺灣墨玉的礦物學特性。在展覽方面，除了由經濟部礦務局、國立東華大學藝術創意產業學系、國立臺灣史前文化博物館、天星礦場、山益礦場、國立臺灣大學地質科學系、經濟部中央地質調查所、中華民國鑛業協進會、臺灣玉發展協會等單位協助提供展品與影音資料參展外，值得一提的是本展策展人之一，知名的金工藝術家，也是東華大學藝術創意產業學系林淑雅教授的玉石創作，利用臺灣墨玉與臺灣玉，參酌古玉石器物形制，轉化、演繹與結合異材質創作玉，開發成為具時尚感的精美金玉首飾與文創飾品併同展示。

當然特別要感謝盧復順、曾保忠、吳照明、張瑞麟、李毓和、林書弘等專家學者，亦提供其精美礦石典藏參展，使得展覽能更加圓滿，在此一併致謝。

本書的出版是作者集結近年來的有關臺灣墨玉的研究成果及金工創作，藉由舉辦展覽的方式呈現，為臺灣玉石留下一頁見證。

作者
國立臺灣博物館研究組副研究員方建能博士
國立東華大學藝術創意產業學系助理教授林淑雅

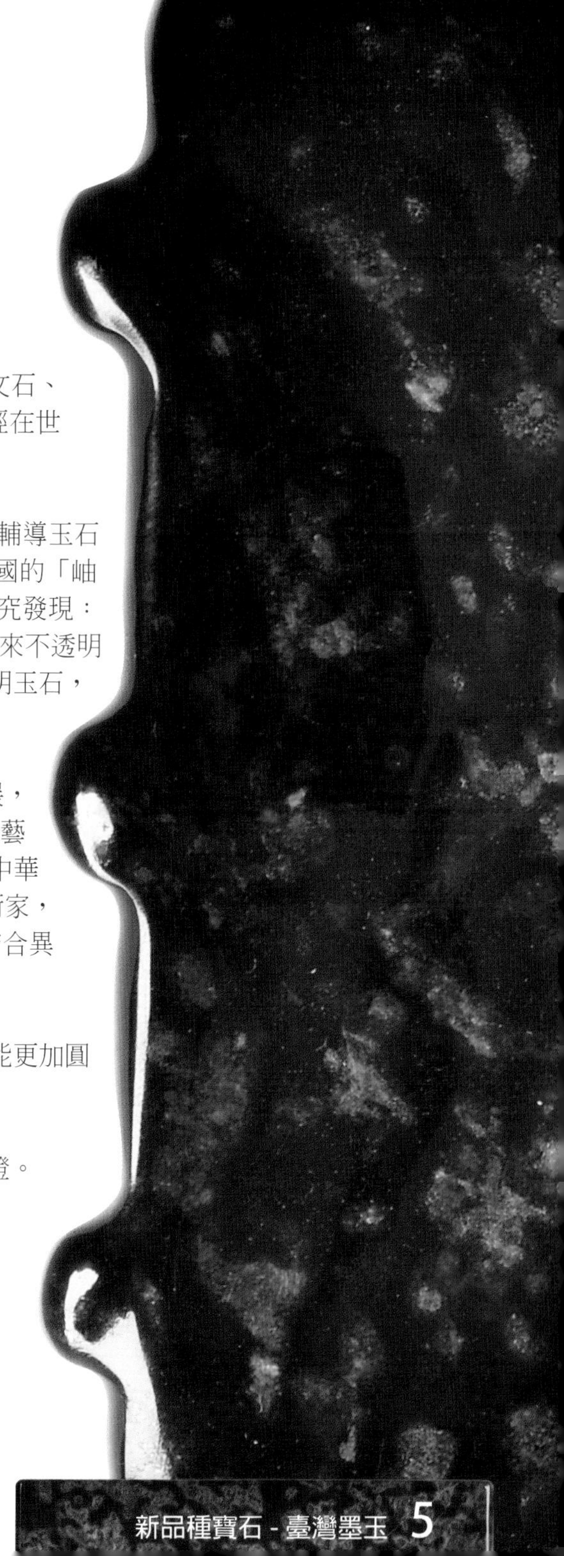

緣起

Prologue

臺灣花蓮至臺東間之花東縱谷，為菲律賓海板塊與歐亞板塊的隱沒碰撞帶，在此地區的岩石產生變質、火成及熱液蝕變等各種不同的地質作用，形成了玫瑰石、臺灣玉、臺灣藍寶、紫玉髓等玉石，及金銅礦、石棉礦、大理岩礦和蛇紋岩礦等礦床，所以花東地區向有「地質天堂」、「玉石故鄉」之美譽。

近年來臺灣的玉石業者在經濟部礦務局的協助下，開發出一種新品種黑色的玉石 - 市場上稱為「臺灣墨玉」，此種玉石常與世界聞名的臺灣玉共生，因其蘊藏量相當驚人，極具開發的潛能。

臺博館特別以「臺灣墨玉」為主題，由「玉」的文化與歷史談起，進而介紹玉石的分類，臺灣玉及臺灣墨玉的礦物學特性，希望帶領國人回顧已有高知名度的臺灣玉，及認識臺灣墨玉 - 這種極具潛力的臺灣本土產玉石。

As a colliding zone between Philippine Sea Plate and Eurasian Plate, the East Rift Valley straddling Hualien and Taitung has been known as a "geological paradise" and the "home of jade stones" due to the metamorphism, pyrogenesis and hydrothermal alteration. Rocks in the region being rich in gold, copper, asbestos, marble, serpentine, and gemstones like rhodonite, Taiwan jade, Taiwan sapphire, and purple chalcedony.

With the efforts of Taiwanese jade suppliers and the assistance from the MOEA (Ministry of Economic Affairs) Bureau of Mine, a unique ink-black jade has been developed. This "Taiwan black jade," as it is known in the market, and is generally associated with the world-renowned Taiwan jade in amazingly rich amount, reserves a great potential of exploitation and development.

Planning the special exhibition on "Taiwan Black Jade," National Taiwan Museum (NTM) starts with the history and cultural significance of jades in China, presents a panoramic introduction of jades by category, and highlights the mineral properties of both Taiwan jades and Taiwan black jade so as to invite visitors to join its celebration for the reputed Taiwan jade and to develop an intimate understanding of Taiwan black jades.

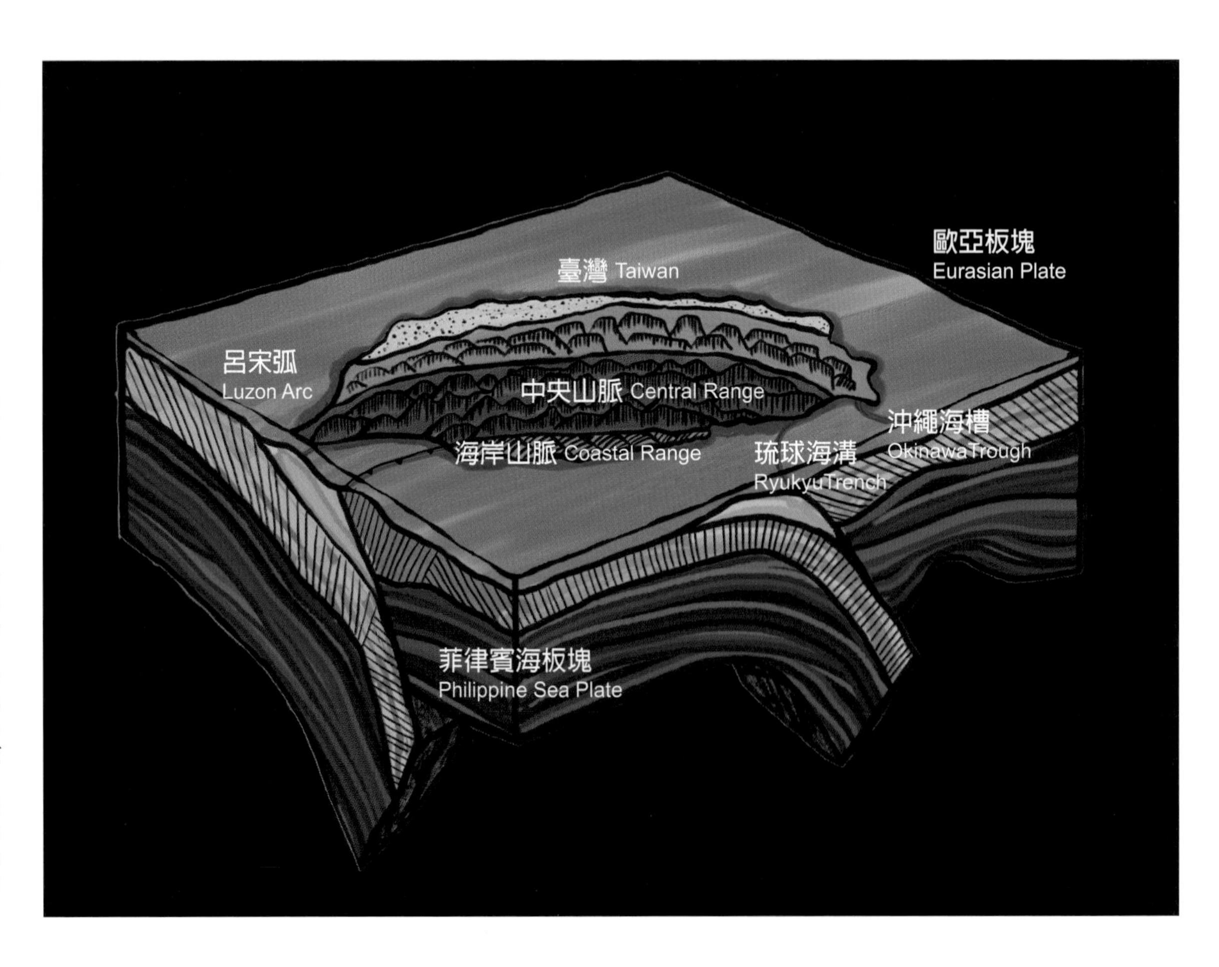

第一章：玉文化與歷史

Chapter 1 : Jade Culture and History

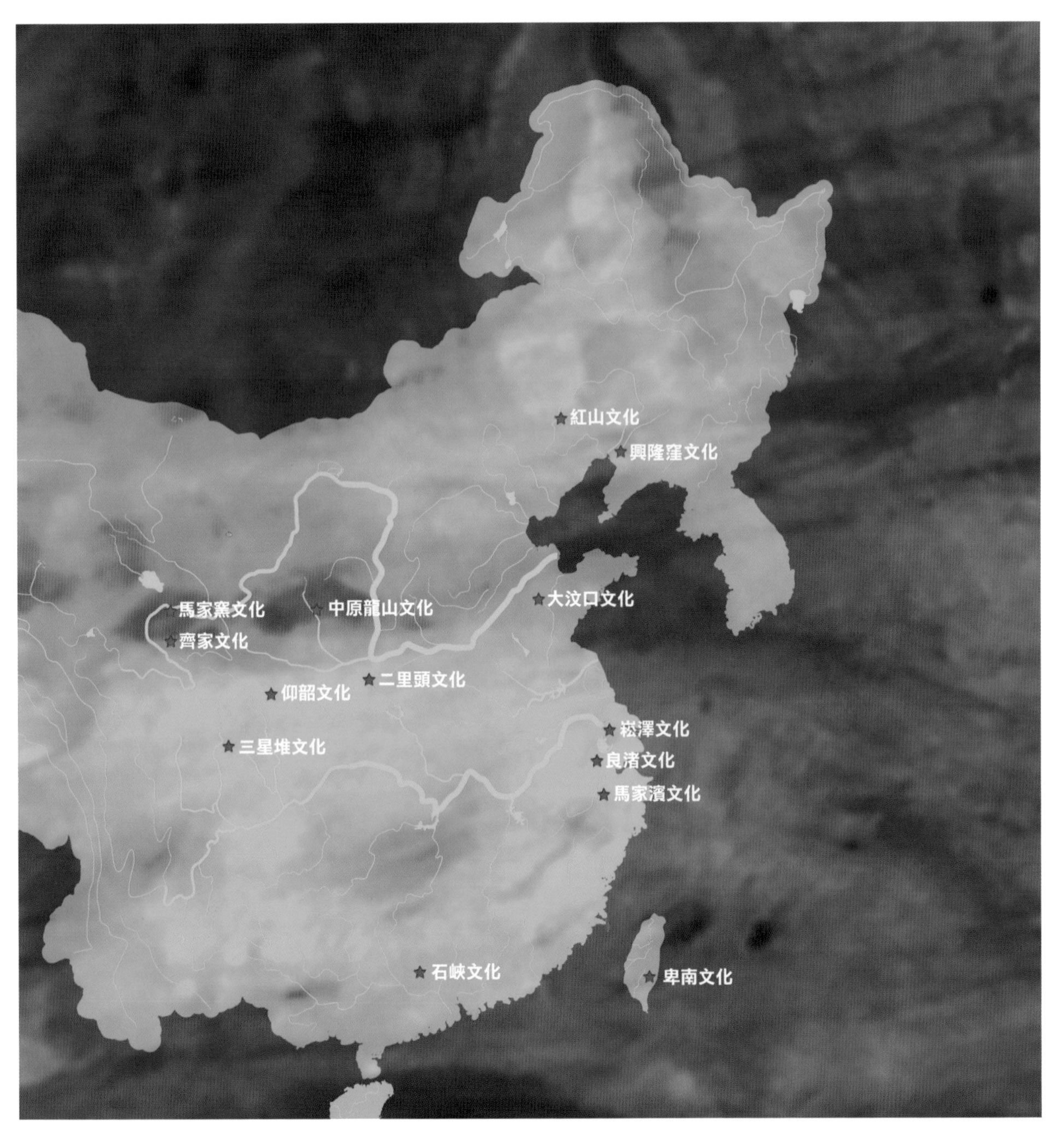

玉文化與歷史

Jade Culture and History

玉在歷史悠久而淵遠流長的八千年中華文化中，是最具代表性的物質，也是國人最喜愛珍藏與佩帶的寶石之一。中華民族的祖先早在八千年前，就已製造出非常精美的玉器。

In the 8,000 years of Chinese history, jade has remained one of the most representative, the most treasured, and the most commonly worn gemstones. Debut in the Chinese culture of delicate and exquisite jadeware can be traced back to as early as 8,000 years ago.

史前時代的玉文化

Prehistoric Jade Culture

玉在中華文化早期就是重要的禮器，在進行祭祀與重要典禮時常需要使用玉器。在史前的新石器時代文化：紅山、龍山、良渚及卑南等都是重要玉文化典型代表。

Jade has been an essential sacrificial ware since the nascence of Chinese culture, registering prominent presence in religious and ancestor worship as well as in major ceremonies as early as in the prehistoric Neolithic period. The Hongshan, Longshan, Liangzhu, and Peinan cultures in particular are known for their opulent jade artifacts.

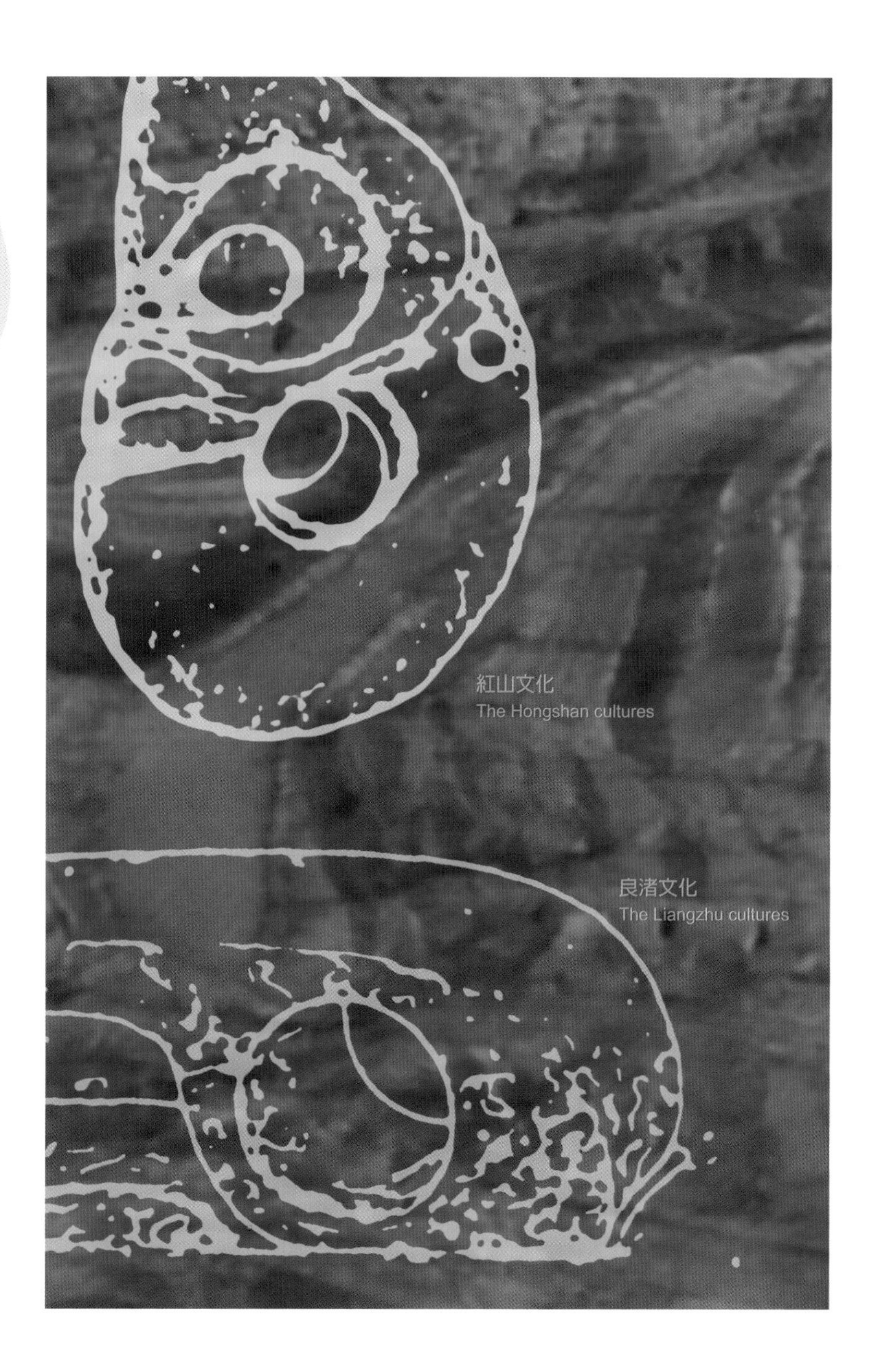

卑南文化出土玉石

Jadeware Unearthed in the Peinan Site

以臺東市卑南遺址為代表，屬於新石器時代晚期文化的「卑南文化」，年代約距今3,500-2,300年間，為臺灣玉器使用最為頻繁的文化體系。

卑南文化玉器器型包括玉珠、玉玦、玉環、玉管、玉錛鑿、玉斧、玉扣、玉棒、玉矛鏃、玉墜等多種，製工之細、種類之多，令人讚嘆。

Dated back to 2,300~3,500 years ago and best represented by the Peinan Site excavated in Taitung City, the Peinan Culture marks a cultural system in Taiwan featuring the most frequent use of jadeware. The jadewares unearthed in the Peinan Site include an amazingly wide spectrum of gaspingly well-crafted jade artifacts, notably jade beads, earnings, bracelets, tubes, chisels, axes, buckles, sticks, spears, arrowheads, and pendants.

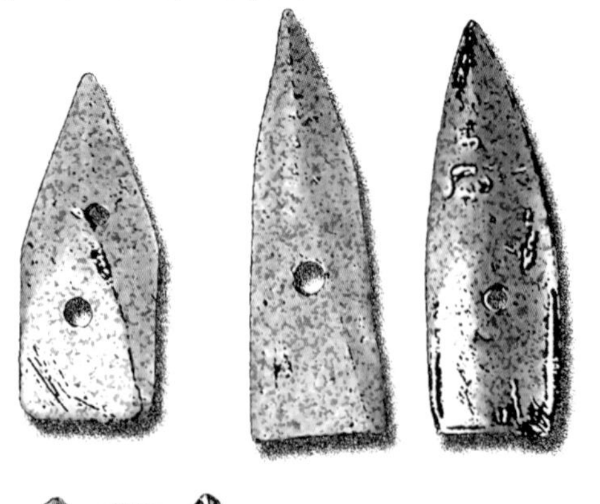

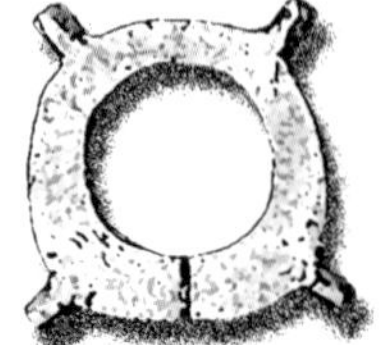

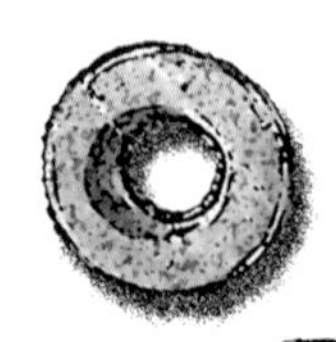

歷史時代的玉文化

Roles and Significance of Jade in the Historical Periods

進入歷史時代以後，透過古籍文字的記載，讓我們更了解玉文化的重要性。

As human history entered the historical periods when ancient written records began to be kept, people were granted a better understanding of the cultural significance of jade in China.

《周禮注疏》：
「以玉作六器。以禮天地四方。以蒼璧禮天。以黃琮禮地。以青圭禮東方。以赤璋禮南方。以白琥禮西方。以玄璜禮北方。」

在周朝時，祭祀天地四方之神的禮器有六種：蒼璧、黃琮、青圭、赤璋、白琥、玄璜。用蒼璧禮天、黃琮禮地（土）、青珪禮東方（木）、赤璋禮南方（火）、白琥禮西方（金）、玄璜禮北方（水）。

In the Zhou Dynasty, the sacrificial ware used to worship the gods of heaven, earth, and the four directions (or "elements") included the so-called six ritual (ceremonial) jades: blue bi (disc) for heaven, yellow cong (pillar) for earth, green gui (tablet) for the east (wood), red zhang (blade) for the south (fire), white hu (tiger-shaped ornament) for the west (gold), and black huang (semi-annular pendant) for the north (water).

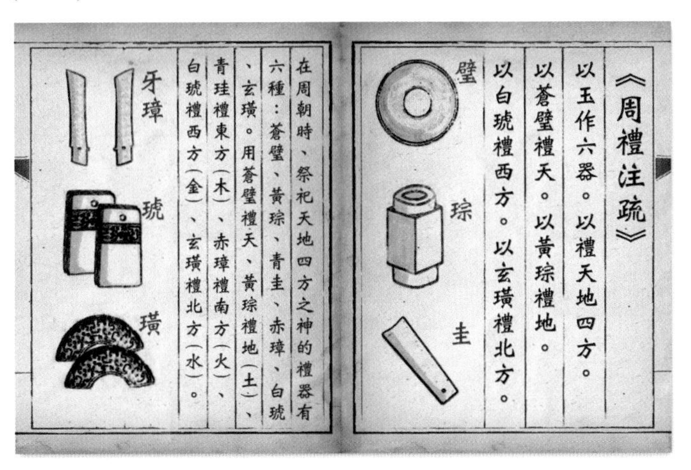
《周禮注疏》
以玉作六器。以禮天地四方。
以蒼璧禮天。以黃琮禮地。
以白琥禮西方。以玄璜禮北方。

在周朝時、祭祀天地四方之神的禮器有六種：蒼璧、黃琮、青圭、赤璋、白琥、玄璜。用蒼璧禮天、黃琮禮地(土)、青珪禮東方(木)、赤璋禮南方(火)、白琥禮西方(金)、玄璜禮北方(水)。

《周禮冬官考工記》：
「玉多則重，石多則輕」；
「玉方寸重七兩，石方寸重六兩」

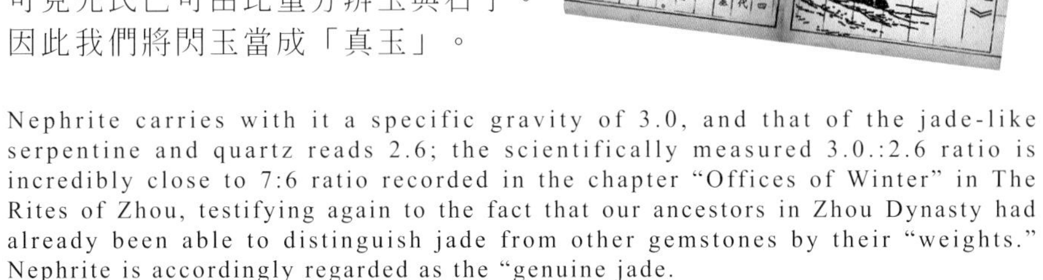
《周禮冬官考工記》
玉多則重，
石多則輕。
玉方寸重七兩，
石方寸重六兩。

閃玉的比重為3.0，一般仿玉的玉石如蛇紋石、石英等比重約2.6，兩者的比例和古書所言「七六比」一致，可見先民已可由比重分辨玉與石了。因此我們將閃玉當成「真玉」。

《周禮冬官考工記》
天子用全，
上公用龍，
侯用瓚，
伯用將。

閃玉的比重為3.0，一般仿玉的玉石如蛇紋石、石英等比重約2.6，兩者的比例和古書所言「七六比」一致，可見先民已可由比重分辨玉與石了。因此我們將閃玉當成「真玉」。

Nephrite carries with it a specific gravity of 3.0, and that of the jade-like serpentine and quartz reads 2.6; the scientifically measured 3.0.:2.6 ratio is incredibly close to 7:6 ratio recorded in the chapter "Offices of Winter" in The Rites of Zhou, testifying again to the fact that our ancestors in Zhou Dynasty had already been able to distinguish jade from other gemstones by their "weights." Nephrite is accordingly regarded as the "genuine jade.

《周禮冬官考工記》：
「天子用全，上公用龍，侯用瓚，伯用將」

其中「全」代表全部是玉，「龍」代表四玉一石，「瓚」代表三玉二石，「將」代表玉石參半。在歷年出土的古代帝王墳墓中，陪葬的玉器全屬於閃玉的材質，而一般王公諸侯的陪葬器，部分為閃玉，部分為其他礦物或岩石，例如蛇紋石或石英等。這說明了玉在古代的尊貴地位，只有帝王才可以享有全部以閃玉為陪葬品的尊榮。有階級層次之別。這也顯示此時已有能力分辨玉與石。

According to the chapter "Offices of Winter" in The Rites of Zhou, only an emperor was entitled to use all jade artifacts as his funerary objects. For an archduke or a duke, the funerary objects were supposed to include 80% jade artifacts and 20% stone artifacts.
For a lord of a lower rank, the funerary objects should be 60% jade artifacts and 40% stone artifacts. For a general, the funerary objects should consist of 50% jade and 50% stone artifacts. All the known funerary objects unearthed from imperial tombs in China are jade (nephrite) artifacts, while the funerary objects of other members of the nobility are a mixture of nephrite artifacts and rock/mineral (such as quartz and serpentine). These archeological facts not only speak eloquently of the privileged status of jade in ancient China but also indicate the ability of Zhou people in telling the difference between jade and other gemstones.

《說文解字》：
「玉，石之美。有五德：潤澤以溫，仁之方也；䚡理自外，可以知中，義之方也；其聲舒揚，專以遠聞，智之方也；不撓而折，勇之方也；銳廉而不忮，絜之方也。」

根據東漢許慎記載，只要是具有美麗外觀的礦物與岩石皆可廣泛稱為「玉」。因此，如閃玉、蛇紋石、玉髓、瑪瑙、琥珀、水晶、青金石等礦石，都曾經被稱為「玉」。
但許慎也描述玉具備的五種特性與意義：「仁」表示玉具有潤澤良好的光澤和透明度，代表善施恩澤，富有仁愛之心。「義」表示玉具有較高的透明度，從外部可以看出來其內部具有的紋理，代表有竭盡忠義之心。「智」表示玉結構細膩緻密少裂痕，敲擊時聲音清脆而遠揚，代表明有智慧並傳達給四周的人。「勇」表示玉具有較高的硬度與韌性，代表具有極高的勇氣。「勇」表明玉具有超人的勇氣；有斷口但邊緣卻不鋒利，「絜」代表玉自身廉潔、自我約束卻並不傷害他人。

In Shuowen Jiezi (literally "Explaining and Analyzing Characters"), the oldest Chinese character dictionary compiled by Xu Sheng in the later Han period, all the minerals and rocks displaying a beautiful appearance can be termed as "jade." Therefore, nephrite, serpentine, chalcedony, agate, amber, crystal, and lapis lazuli have been called "jade" one time or another in Chinese history.
According to Xu Sheng, jade embodies five virtues exalted in Chinese culture:
"Jen," reflected by the graceful luster and tender warmth of jade, refers to benevolence and philanthropy.
"Yi," represented by the fine transparency of jade that allows one to appreciate its inner texture, refers to rectitude and righteousness.
"Zhi," symbolized by the clear, crispy sound issued by the tapping of genuine jade and its delicate and dense structure, refers to wisdom ready for sharing and spreading.
"Yung," mirrored by the exceptional toughness of jade, refers to courage and strength.
"Lian," illustrated by the absence of jagged or rough edges in most jade artifacts, refers to self-discipline and integrity.

《禮記・玉藻》：
「古之君子必佩玉。右徵角，左宮羽」

《禮記・玉藻》是孔子說玉的專章：商周時的有修養的士大夫階層都會佩戴玉，以體現出周禮的威嚴與莊重。行走時右邊的佩玉發出合於五音中徵、角的聲音，左邊的佩玉發出合于宮、羽的聲音。

To quote Confucius in Yuzao, the chapter of Liji (The Book of Rites) focusing on the jade-bead pendants of the royal cap, cultured scholar-officials in Shang and Zhou dynasties wore jade ornaments to epitomize their respect for rituals and etiquette. The jade ornaments were worn in such a particular manner that those on the right side would produce the "zhi" and "jiao" sounds and those on the left side the "gong" and "yu" sounds of the Chinese "five tones" when a scholar-official was walking.

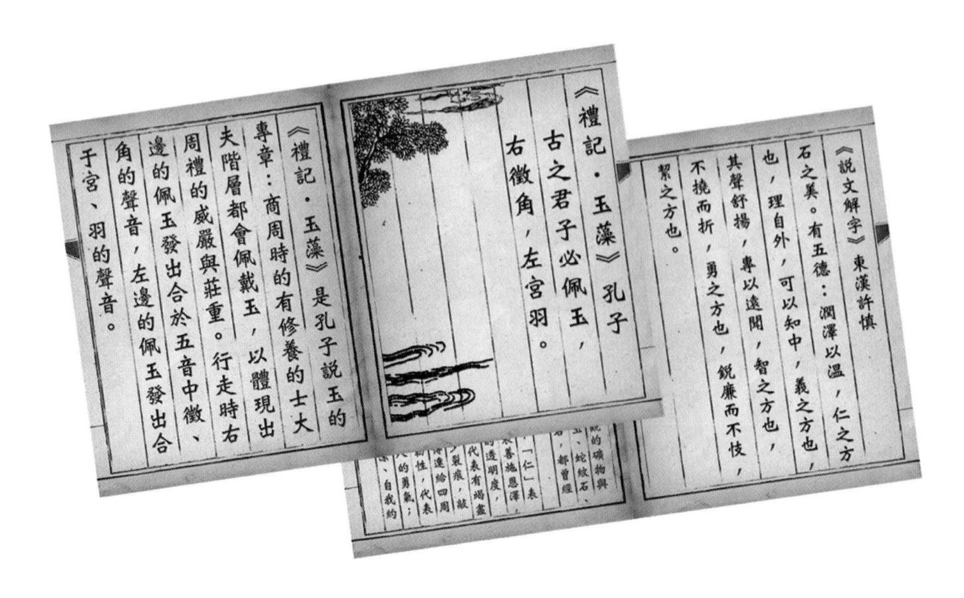
《禮記・玉藻》孔子
古之君子必佩玉，
右徵角，左宮羽。

《禮記・玉藻》是孔子說玉的
專章：商周時的有修養的士大
夫階層都會佩戴玉，以體現出
周禮的威嚴與莊重。行走時右
邊的佩玉發出合於五音中徵、
角的聲音，左邊的佩玉發出合
于宮、羽的聲音。

《說文解字》東漢許慎
石之美。有五德：潤澤以溫，仁之方
也，理自外，可以知中，義之方也，
其聲舒揚，專以遠聞，智之方也，
不撓而折，勇之方也，銳廉而不忮，
絜之方也。

文字中的玉

Jade-related Words

中國文字中以玉為邊者多達三百字，除了代表禮器、玉德之用途外，玉也是官階和爵位的表徵，而且還有各種符節、簡冊、刀斧、樂器、服飾等意義，可見玉已成為中華文化重要的一環了。

There are over 300 Chinese words with the "jade" radical on either side. In addition to ceremonial vessels and jade-associated virtues, these words also used to denote orders of nobility, officer ranks, emblems and symbols of authority, ancient books made of bamboo or wooden slips, swords and axes, musical instruments, clothing, and personal adornments. Its extensive presence in Chinese characters betokens the rich significance of jade in Chinese culture.

瑁瑃瑄瑅瑆瑇瑈瑨瑊瑌瑍瑎瑏瑜瑒瑓瑔瑖瑘瑝
玉玗玖璵玟玠玢玥玦玨玩玲珉玷玭珀珂
珩玻珊珠琢琮琵琺
玌玍玎玏玐玒玓玔玗玘瑒玜玝玞玡玣玤珫玴玵玶
琇琉琍璡璉瑣琊琚琛琤琥琦琨琪琬琯琰
玸玹玽玾玿珁珃珄珅珆珇瑨珋珌珎珒珓珔珕珧珗
琲琳琴琶琿瑁瑋瑑瑕瑗瑚瑛瑞瑟瑤瑣瑤
珘珚珛珜珝珟珡珢珣珤珦珨珫珬珯珱珳珴珵珶
珸珹珺珻珼珽琀琁琂琄琈琋琌琑琒琓珷琔琕琖琗
瑩瑭瑯瑰瑱瑲璦瑾璀璆璯璡瓔璜璞璟璠
璡璣璐璘璦璨璩璫璯璵璽璿瓊瓔瓖瓚
瑨琙琜琝琟琠琡琣琧琩琫琭琱琎琸琹琻琞琽琾瑀
瑠瑯瑢瑥瑦瑧瑨瑫瑬瑮瑳瑴瑵瑸瑹瑺瑻瑼瑽瑿
瑜瑙瑪璃璋璧環瓏
璁璂璄璅璈璊璌璍璣璑璒璓璔璕璖璗璙璚璛璝
珈珍珙珞珥珧珪班珮璫琿頊珽現球琅理
璏璢璥璪璬璭璮璱璲璳璴璶璸璹璺璻璼璾瓀瓁
瓂璬瓃瓄瓅瓆瓇瓈瓋瓌瓍瓎瓐瓑瓓瓕瓗瓘瓙瓛

何謂「玉」？
What is "Jade"?

「玉」的古今定義，到底什麼是「玉」呢？古今所認知的「玉」有何差異呢？讓我們一起來認識吧！

What is "Jade"? Definitions of Jade: Ancient and Modern. What exactly, after the above introduction, is "jade"? Are there any differences between the ancient and modern perceptions of "jade"? It is time to find out.

古代「玉」的定義
Ancient Definition of Jade

透過古書的記載，我們了解在清朝以前，被認定為真正的「玉」指的是閃玉而言，所以閃玉又稱「真玉」；除了閃玉以外，其他具漂亮外觀的美石，我們稱為「似玉」、「非玉」、彩石或假玉。
現今較受喜愛的輝玉，是在清朝後期才開始流行。

Based on the records of ancient documents, we understand that, prior to the Qing Dynasty, the "genuine" jade referred to nephrite. In addition to nephrite, there were other exquisite and beautiful "jade-like," "non-jade," or "pseudo-jade" gemstones. The pyroxene jade that is fairly well favored in modern time did not gain its popularity until late Qing Dynasty.

現代「玉」的定義
Modern Definition of Jade

現代寶石學家認定的「玉」，是指閃玉與輝玉兩種而言。閃玉的材質是屬於角閃石類礦物；而輝玉的材質則是屬於輝石類礦物。

For modern gemologists, "jade" refers to either nephrite or jadeite. Nephrite is a fine-grained tremolite or actinolite , and jadeite a massive isomorphous mixture of acmite and diopside.

玉石小博士 - 閃玉與輝玉名稱的由來
Greeting from Dr. Jade: Origins of "Nephrite" and "Jadeite"

閃玉俗稱「軟玉」、「和闐玉」、「新疆玉」，輝玉俗稱「硬玉」、「緬甸玉」、「翡翠」。事實上，閃玉與輝玉的英文名稱 nephrite 和 jadeite，完全沒有軟和硬之意。但因為早期有些礦物學書籍記載著 Nephrite 的硬度為 5.5-6.5，jadeite 的硬度為 6-7，日本人早期翻譯這兩個名詞時，便分別用「軟」、「硬」予以命名。
然而根據新的研究顯示，所謂「軟玉」的閃玉的硬度是 5.5-7，甚而有高於 7 者，而所謂「硬玉」的輝玉硬度是 6-7，兩者硬度相當接近，以軟硬命名會給人錯覺，以為軟玉真的很軟，這顯然不適當。
由於以上的理由，已退休的國立臺灣大學地質科學系譚立平教授，根據兩者礦物的種類加以命名：nephrite 的主要成分為角閃石類（amphibole）礦物，jadeite 為輝石類（pyroxene）礦物，建議 nephrite 稱為閃玉，jadeite 稱為輝玉。目前閃玉與輝玉的中文名稱，已在國內外廣泛被引用。

Nephrite is also known as "soft jade," "Heitian jade," and "Xinjian jade," while jadeite is commonly referred to as "hard jade," "Burma jade," and "emerald."The fact that in their English forms, the names "nephrite" and "jadeite" are by no means associated with either softness or hardness. However, it is stated in some early mineralogy documents that the hardness of nephrite falls in the range of 5.5~6.5, and that of jadeite 6~7. As a result, the two terms – nephrite and jadeite - were first translated into Japanese using "soft" and "hard" for distinction.
Subsequent studies indicate that the hardness of the so-called "soft" nephrite reaches 5.5~7, and nephrites with a hardness surpassing 7 have been detected. The fact that nephrite and jadeite are neck and neck in terms of hardness exposes the obvious incongruity in referring nephrite as "soft jade" and jadeite "hard jade" as it is very likely to cause the misunderstanding that nephrite is indeed "soft" in nature.
Keenly aware of the problem, Dr. Tan, Li-Ping, a respected and retired professor at the Department of Geosciences, National Taiwan University, proposed to re-translate nephrite and jadeite into Chinese from a mineralogical standpoint. Nephrite, as an amphibole ("Shan stone" in Chinese, literally "radiant stone"); is accordingly renamed into "shan yu" ("radiant jade") in Chinese. Jadeite, as a pyroxene ("Hui stone" in Chinese, literally "splendid stone"), is called "hui yu" ("splendid jade"). Both Chinese terms, "shan yu" for nephrite and "hui yu" for jadeite," have since been used extensively.

第二章：臺灣玉

Chapter 2 : Taiwan Jade

世界聞名的臺灣玉
The World-renowned Taiwan Jade

臺灣生產的閃玉大量開採與加工製造源起民國 50 年代，六十年代為全盛時期，歷時十餘年，每年平均開採超過一千五百公噸，全臺加工廠超過八百家，直接或間接從事玉石業人員高達五萬人以上。
當時，臺灣加工生產的閃玉玉石數量佔當時全世界約八成左右，每年平均銷售金額約新臺幣五十億元，而使臺灣成為世界珠寶加工及出口之王國。因此臺灣生產的綠色閃玉，以「臺灣玉」為名，揚名全世界。
世界各地出產閃玉的國家很多，不單只有臺灣，在中國、美國、加拿大、俄羅斯、澳洲、韓國等亦有產出。

Mass exploitation and processing of nephrite in Taiwan started in the 1960s and reached its peak in the 1970s. During the two decades, more than 1,500 tons of nephrites were exploited and processed every year, supporting an industry with over 800 factories and 50,000 workers.
Back then, nephrite artifacts processed in Taiwan accounted for approximately 80% of the global supply with an annual sales revenue exceeding NT$5 billion, making Taiwan the world's leading processor and exporter of gemstones. The green nephrites from Taiwan are known around the world as "Taiwan jade."
In addition to Taiwan, there are many other countries producing and processing nephrites, notably Australia, Canada, China, Korea, Russia, and the United States.

工人去雜石

臺灣玉原石

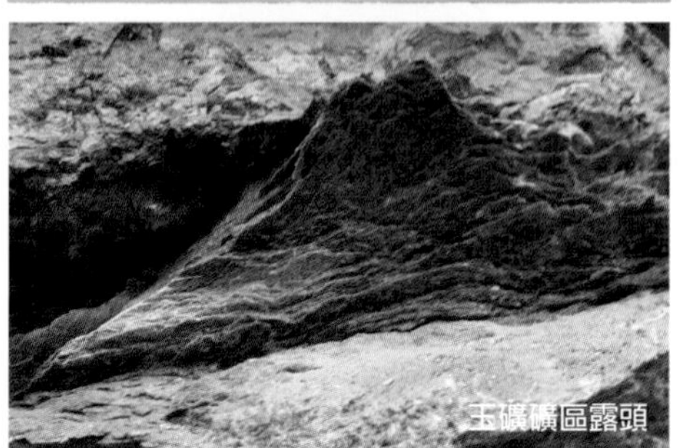
玉礦礦區露頭

礦場開挖玉礦礦石

玉礦坑口

臺灣玉的分類

Types of Taiwan Jade

國立臺灣大學地質科學系譚立平教授在 1978 年《國科會專刊》第一號中，將臺灣玉分成三大類：

(一) 普通閃玉：或簡稱普通玉，是最常見的閃玉，有玻璃光澤，中度透明，其組成礦物的結晶粒度為 40 至 150 微米。

(二) 貓眼閃玉：或稱貓眼玉、臺灣貓眼，為呈現貓眼現象的臺灣玉，市場也稱之為貓眼石。結晶粒度有時超過 1000 微米。

(三) 蠟光閃玉：指光澤似蠟、不透明至半透明的閃玉，因結晶顆粒細微，通常小於 15 微米，光線會產生漫射而發出蠟質光澤。

According to a paper by Professor Tan, Li-Ping published in the first NSC (National Science Council) Special Publication in 1978, there are three main types of Taiwan jade:
1. Common Nephrite: Also called "common jade," this most available nephrite is marked with glassy luster and translucence. The grain size of its major minerals falls in the range of 40~150 μm.
2. Cat's-eye Nephrite: Also referred to as "cat's-eye jade" or "Taiwan cat's eye," this type of nephrite gains the name thanks to its glitter pattern that looks like a cat's eye. When people are talking about "cat's eye stones," they are in fact referring to gemstones processed from cat's-eye nephrite.The grain size of its major minerals is longer than 1000μm.
3. Waxy Nephrite: Known for its waxy luster, this slightly transparent or non-translucent nephrite has a size smaller than 15 μm. The waxy luster is produced as the light diffuses after entering the sophisticated, fine-grained nature of this waxy jade.

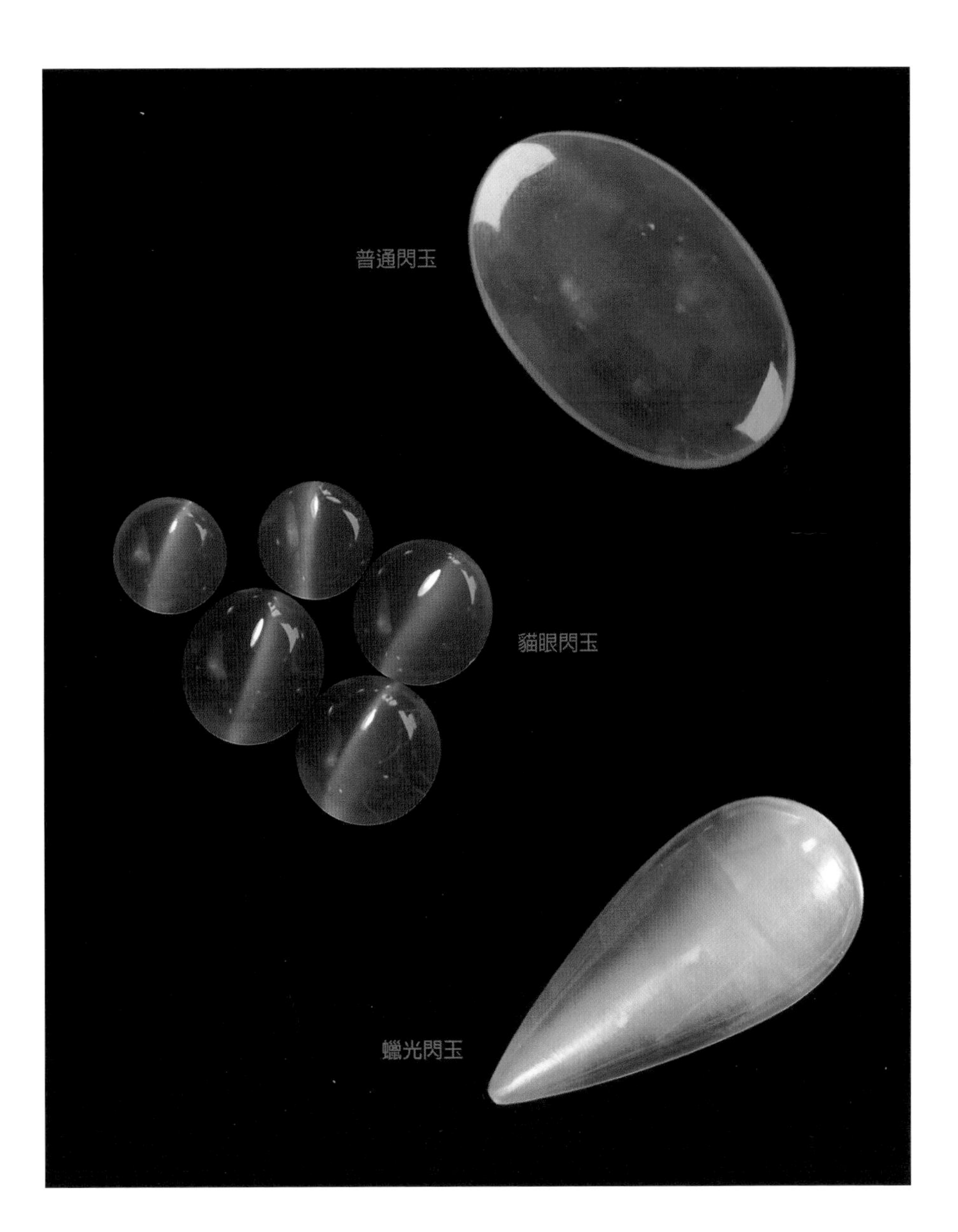

臺灣玉的產狀與礦體分佈

Distribution of Taiwan Jade Deposits and Shapes of Ore Veins

臺灣玉產在臺灣最古老的岩層 - 大南澳片岩中的蛇紋岩和片岩的接觸帶中，年代可溯及古生代二疊紀到中生代間，是由超基性火成岩變成之蛇紋岩再經熱液換質作用而成。

臺灣玉在地層中，礦脈一般呈不規則細脈狀或扁豆狀，厚度以 0.1-0.5 公尺為主，局部可達 1.5-2.0 公尺，長 20-30 公尺。

Most of the deposits of Taiwan Jade are located in the contact zones between the serpentinites and the schists in the Tananao Schist whose formation can be traced back to the period from the Permian Period of the Paleozoic Era to the Mesozoic Era. The jade deposits were formed by ultra-basic igneous rocks metamorphosed into serpentinites that in turn experienced hydrothermal alterations.

The ore veins of Taiwan jade appear to be irregular or lenticular fine veins with a thickness of 0.1~0.5 meters (with the maximum thickness reaching 1.5~2.0 meters) and a length of 20~30 meters.

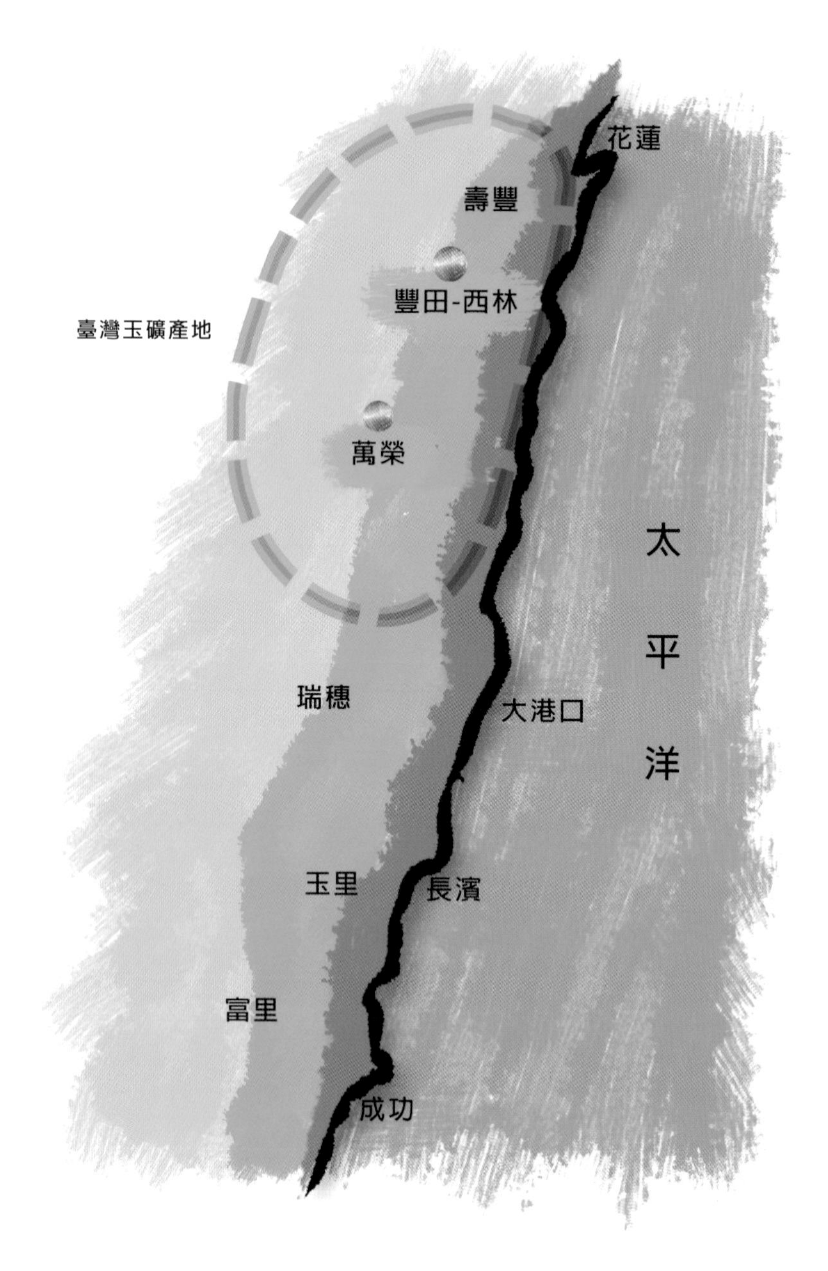

臺灣玉的礦物特性

Mineral Properties of Taiwan Jade

臺灣玉主要是由漂亮且半透明的富含鎂之透閃石或富含鐵之陽起石的角閃石類礦物所組成，玻璃光澤，條痕白色，硬度5.5-7，比重3.0-3.1。臺灣玉常與石棉、滑石、鉻鐵礦、黃銅礦、鉻尖晶石、鈣鋁榴石等礦物共生。

Taiwan jade, mainly composed of the beautiful, semi-transparent, magnesium-rich tremolite or the iron-rich actionlite, reveals glassy luster and white streaks and reports a hardness of 5.5~7 and a specific gravity of 3.0~3.1. Taiwan jade often contains chromite, asbestos, talc, chalcopyrite, picotite, and grossularite.

鉻鐵礦

化學成分 $FeCr_2O_4$，硬度5.5，比重4.3-4.6，金屬光澤，鐵黑至褐黑色，條痕暗褐色至灰色，不透明，磁性弱。

Chromite

Chemical composition $FeCr_2O_4$; hardness 5.5; specific gravity 4.3~4.6; metallic luster; colors ranging from iron black to dark brown; streaks appeared to be dark brown or gray; non-transparent, and magnetism weak.

石棉

化學成分 $Mg_6Si_4O_{10}(OH)_8$，硬度4-6，比重2.5-2.6，具纖維狀外觀，顏色變化多。

Asbestos

Chemical composition $Mg_6Si_4O_{10}(OH)_8$; hardness 4~6; specific gravity 2.5~2.6; with a fibrous appearance, and in various colors.

滑石

化學成分 $Mg_3Si_4O_{10}(OH)_2$，硬度1，比重2.8，半透明，有珍珠光澤，無色至淡綠色，條痕為無色，有滑膩之觸感。

Talc

Chemical composition $Mg_3Si_4O_{10}(OH)_2$; hardness 1; specific gravity 2.8; pearl-like luster; colors ranging from achromatic to light green; streaks appeared to be achromatic and with a smooth, slippery touch.

臺灣玉原石

黃銅礦

化學成分 $CuFeS_2$，硬度3.5-4，比重4.1-4.3，金屬光澤，銅黃色，條痕爲綠黑色，不透明，具有導電性。

Chalcopyrite

Chemical composition $CuFeS_2$; hardness 3.5~4; specific gravity 4.1~4.3; metallic luster; bronze yellow in color; streaks appeared to be dark green; non-transparent, and electrically conductive.

鉻尖晶石

化學成分 $(Mg,Fe_2)(Al,Cr)_2O_4$，硬度7.5-8，比重3.6-4.1，玻璃光澤，黃到綠褐色，條痕爲白色，透明至半透明。

Picotite

Chemical composition $(Mg, Fe_2)(Al,Cr)_2O_4$; hardness 7.6~8; specific gravity 3.6~4.1; glassy luster; colors ranging from yellow to brownish-green; streaks appeared to be white and semi-transparent or transparent.

鈣鋁榴石

化學式爲 $Ca_3Al_2(SiO_4)_3$，硬度6.5-7，比重3.4-3.6，玻璃光澤，黃、褐、紅或綠色，條痕爲白色，透明至不透明。

Grossularite

Chemical composition $Ca_3Al_2(SiO_4)_3$; hardness 6.5~7; specific gravity 3.4~3.6; glassy luster; yellow, brown, red or green in color; streaks appeared to be white and non-transparent, semi-transparent, or transparent.

最早切割的臺灣玉標本

The Earliest Dissecting Sample of Taiwan Jade

臺灣玉的發現源自於 1956 年，當年臺南工學院（成功大學前身）礦冶系廖學誠先生於花蓮豐田的中國石礦公司礦區中首先採集，並鑑定證實為「閃玉」。
本展品即為 1957 年於中國石礦公司礦區所採集，原重 55 公斤，經多次切割而成，由臺灣大學地質科學系譚立平教授（已退休）提供。

Taiwan jade was first discovered and identified as nephrite in 1956 in the Fengtien (Hualien) mining district of China Mining Co., Ltd. by Mr. Liao, Hsueh-Cheng from the Department of Mining & Metallurgical Engineering, Tainan College of Engineering (predecessor of today's National Cheng Kung University). The exhibit here is a part of the original Taiwan jade, which weighed 55 kg at the time of its discovery in 1957(kindly provided by Dr. Tan, Li-Ping, retired professor at the NTU Department of Geosciences).

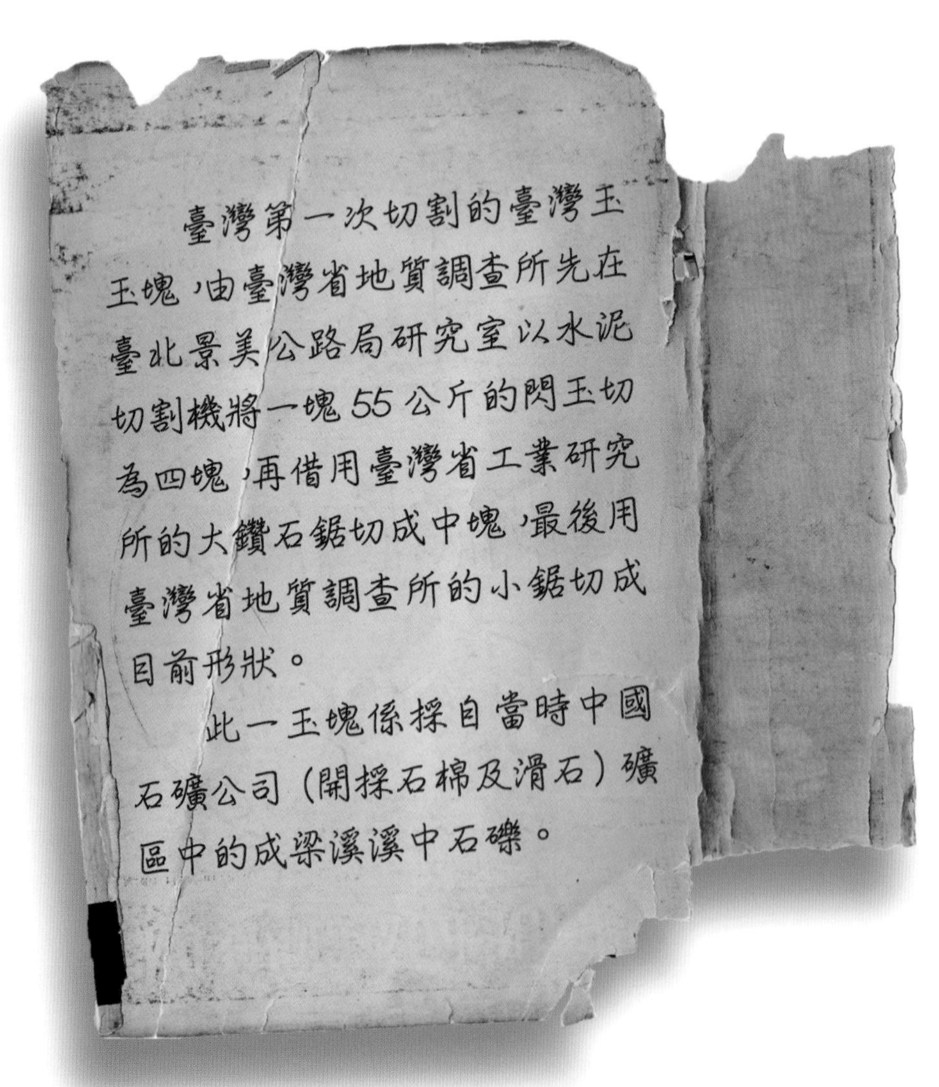

臺灣第一次切割的臺灣玉
玉塊，由臺灣省地質調查所先在
臺北景美公路局研究室以水泥
切割機將一塊 55 公斤的閃玉切
為四塊，再借用臺灣省工業研究
所的大鑽石鋸切成中塊，最後用
臺灣省地質調查所的小鋸切成
目前形狀。

此一玉塊係採自當時中國
石礦公司（開採石棉及滑石）礦
區中的成梁溪溪中石礫。

第三章：臺灣墨玉

Chapter 3 : Taiwan Black Jade

臺灣墨玉的歷史溯源

History of Taiwan Black Jade

臺灣墨玉是一種臺灣產的寶石級黑色的蛇紋石，追溯歷史文獻可發現，這種寶石並不是現在才被發現。早在民國 67 年 - 行政院國家科學委員會出版研究專刊第 1 號 - 臺灣花蓮豐田軟玉礦床之礦物研究報告即指出：「在豐田地區出產的黃綠色半透明的蛇紋石 - 鮑文玉也被開採當作臺灣玉來賣，由於以往因未發現大型礦床賦存，因此並未單獨開採利用，只偶而被魚目混珠當作閃玉來賣」。

Taiwan black jade is a black, gem-class serpentine from Taiwan. Historical documents suggest that Taiwan jade is in fact not a product of recent years. As early as in 1978, a paper, entitled "A Mineralogical Study on the Fengtien Nephrite Deposits in Hualian, Taiwan" and published in the previously mentioned NSC Special Publication No.1, reported that "The yellowish-green and semi-transparent bowenite or nephrite bowen produced in the Feng Tian district has been sold as Taiwan jade. Since the deposits are not large-scale ones, no mass exploitation has been attempted. Only occasionally is the Feng Tian bowenite or nephrite bowen masqueraded as nephrite for sales in the market."

譚立平、王執明、陳其瑞、田沛霖、崔伯銓、俞震甫（1978）行政院國家科學委員會專刊第 1 號 - 臺灣花蓮豐田軟玉礦床之礦物研究報告。
Tan, Li-Ping; Wang, Chih-Ming; Chen, Chi-Chieu; Tien, Pei-Lin; Tsui, Po-Chuan; and Yu, Tzen-Fu. "A Mineralogical Study of the Fengtien Nephrite Deposits in Hualien, Taiwan." NSC Special Publication No.1, 1978.

臺灣墨玉的命名
Naming of Taiwan Black Jade

2010 年國立臺灣博物館副研究員方建能博士進行詳細礦物學研究後，建議稱這種臺灣產的蛇紋石玉為「臺灣墨玉」，以便與世界其他寶石有所區分。

In 2010, after detailed mineralogical studies, associate researcher Dr. Fang, Jian-Neng at National Taiwan Museum suggested to name the black gem-class serpentine as "Taiwan black jade" to mark its unique distinction from the other similar gemstones in the world.

方建能、曾保忠、陳韻婕 (2011) 臺灣墨玉 - 臺灣新品種的潛力寶石。臺灣博物季刊，30（1）：82-85。
方建能、陳韻婕、余炳盛、曾保忠 (2012) 臺灣蛇紋石玉。臺灣鑛業，6（3）：1-8。

Fang, Jian-Neng; Tseng, Pao-Chung; and Chen, Yun-Chie. "Taiwan Black Jade: Potential of a New Breed of Taiwan Gemstone." Quarterly Journal of Taiwan Museum. 30(1): 82-85.
Fang, Jian-Neng; Chen, Yun-Chie; Yu, Bing-Sheng; and Tseng, Pao-Chung. "Taiwan Serpentine Jade." The Taiwan Mining Industry. 6(3): 1-8.

玉石小博士 - 鮑文玉
Greeting from Dr. Jade: On Bowenite (Nephrite Bowen)

鮑文玉是由蛇紋石礦物組成的寶石級玉石。

Bowenite or nephrite bowen refers to a gem-class stone composed mainly of serpentine.

玉石小博士 -「岫岩玉」、「岫玉」
Greeting from Dr. Jade: On Xiuyan Jade

世界上的蛇紋石玉，以產自中國遼寧省岫岩縣者品質最佳，也最富盛名，故稱為「岫岩玉」。岫岩玉由於顏色美觀，硬度低，容易加工，具有耐高溫性和抗腐蝕性，可雕性和透光性好，是優質的玉雕材料。
因為中國遼寧省岫岩縣產的岫岩玉如此的著名，為了區別中國境內所產同屬蛇紋石玉，亦常被冠上「岫玉」的名稱，常見的例如產於祁連酒泉地區，就稱「酒泉玉」和「酒泉岫玉」；產於崑崙山地區的蛇紋石玉，就稱「崑崙岫玉」等。

When it comes to serpentine jade, the one from Xiuyan County of Liaoning Province in China is generally regarded as of the best quality worldwide. Hence the name "Xiuyan jade." Other than its beautiful color, Xiuyan jade is perfect for carving owning to its low hardness, easy processability, high-temperature resistance, robust corrosion resistance, and superior polishing property.
The celebrated fame of Xiuyan jade from Liaoning Province has prompted suppliers of serpentine jade in other regions of China to blend the term "Xiuyan jade" into the names of their products. The serpentine gemstones from Qilian and Jiuquan, for example, are called "Jiuquan jade" or "Jiuquan Xiuyan jade," and those from the Kunlun Mountains district "Kunlun Xiuyan jade.".

Mineral Properties of Taiwan Black Jade

臺灣墨玉主要由半透明的葉蛇紋石、纖蛇紋石與不透明磁鐵礦組成，有時也伴隨少量的蜥蛇紋石、鎂橄欖石、透輝石、白雲石與滑石等礦物。

Taiwan black jade is mainly composed of semi-transparent antigorite, chrysotile and non-transparent magnetite, sometimes blended with lizardite, forsterite, diopside, dolomite, talc, and other minerals in small amounts.

1. 葉蛇紋石
化學成分$Mg_3Si_2O_5(OH)_4$，單斜晶系，硬度3-5，比重2.5-2.6，半透明到不透明，綠色，條痕白色，多呈板狀、葉狀或緻密塊狀，具有油脂或蠟狀光澤。

1. Antigorite
Chemical composition $Mg_3Si_2O_5(OH)_4$; monoclinic system; hardness 3~5; specific gravity 2.5~2.6; semi-transparent or non-transparent; green in color; streaks appears to be white and with greasy or waxy luster.

2. 纖蛇紋石
化學成分$Mg_3Si_2O_5(OH)_4$，斜方晶系，硬度3-5，比重2.5-2.6，半透明到不透明，纖維狀，顏色白色或綠色，條痕白色，絹絲光澤。

2. Chrysotile
Chemical composition $Mg_3Si_2O_5(OH)_4$; orthorhombic system; hardness 3~5; specific gravity 2.5~2.6; semi-transparent or non-transparent; white or green in color; streaks appears to be white and with silky luster.

3. 蜥蛇紋石
化學成分 $Mg_3(Si_2O_5)(OH)_4$，三斜晶系，硬度 3-5，比重 2.6，半透明到不透明，平板狀或細粒塊狀，顏色多為綠至黃色，條痕白色，具油脂至蠟狀光澤。

3. Lizardite
Chemical composition $Mg_3(Si_2O_5)(OH)_4$; triclinic system; hardness 3~5; specific gravity 2.6; semi-transparent or non-transparent; yellow or green in color; streaks appears to be white and with greasy or waxy luster.

4. 磁鐵礦
化學成分 Fe_3O_4，等軸晶系，硬度6，比重5.2，不透明，多八面體與十二面體，粒狀及塊狀，黑色，條痕黑色，金屬光澤。具有強磁性，可以當做自然磁鐵用，又稱磁石。

4. Magnetite
Chemical composition Fe_3O_4; isometric system; hardness 6; specific gravity 5.2; non-transparent; black in color; streaks appears to be white and magnetism strong enough for use as a natural magnet (a.k.a. lodestone).

5. 赤鐵礦
化學成分 Fe2O3，六方晶系，硬度 6.5，比重 5.3，半透明至不透明，土狀及塊狀，紅灰、黑、暗紅色，條痕紅棕色，金屬光澤。

5. Hematite
Chemical composition Fe2O3; hexagonal system; hardness 6.5; specific gravity 5.3; subtranslucent to opaque; earthy and blocky; reddish gray, black, blackish red in color; streaks appeared to be reddish brown; metallic luster.

6. 鎂橄欖石
化學成分為 $(Mg, Fe)_2SiO_4$，斜方晶系，硬度 6.5 -7，比重為 3.3- 4.3，透明至半透明，橄欖綠色，條痕白色，短柱狀或粒狀，具玻璃光澤。

6. Forsterite
Chemical composition $(Mg, Fe)_2SiO_4$; orthorhombic system; hardness 6.5~7; specific gravity 3.3~4.3; transparent or semi-transparent; olive green in color; streaks appears to be white and with glassy luster.

7. 透輝石
化學成分為$CaMgSi_2O_6$，單斜晶系，硬度5.5-6.5，比重3.2 -3.5，透明至半透明，多綠色或褐色，條痕白至灰色，粒狀或是塊狀，具玻璃光澤。

7. Diopside
Chemical composition $CaMgSi_2O_6$; monoclinic system; hardness 5.5~6.5; specific gravity 3.2~3.5; transparent or semi-transparent; mainly green or brown in color; streaks appears to be white and gray and with glassy luster.

8. 白雲石
化學成分 $CaMg(CO_3)_2$，六方晶系，硬度 3.5-4，比重 2.9，透明到半透明，白、灰至淡紅色，條痕白至灰色，晶體多作菱面體，粒狀及緻密塊狀，玻璃光澤。

8. Dolomite
Chemical composition $CaMg(CO_3)_2$; hexagonal system; hardness 3.5~4; specific gravity 2.9; transparent or semi-transparent; mainly white, gray, or light red in color; streaks appears to be white and gray and with glassy luster.

9. 滑石
化學成分為 $Mg_3Si_4O_{10}(OH)_2$，單斜晶系，硬度 1，比重 2.7-2.8，顏色灰至白，條痕白色，半透明，常呈板狀或葉片狀，珍珠至油脂光澤。

9. Talc
Chemical composition $Mg_3Si_4O_{10}(OH)_2$; monoclinic system; hardness 1; specific gravity 2.7~2.8; semi-transparent; gray or white in color; streaks appears to be white; crystals laminar or foliate in shape and with pearl-like or greasy luster.

臺灣墨玉原石

臺灣墨玉的成因
Formation of Taiwan Black Jade

臺灣墨玉如何生成？
How does Taiwan black jade form?

一般而言，地殼深處的超基性岩石或橄欖岩中的橄欖石與輝石礦物，和水作用之後，會轉變成蛇紋石類礦物與磁鐵礦，形成蛇紋岩，這就是所謂的蛇紋岩化作用。
如果蛇紋岩中的蛇紋石類礦物呈現透明或半透明，加上不透明的磁鐵礦，就形成了臺灣墨玉。

Generally speaking, if ultrabasic rocks or peridotite in the deep Earth's crust react with water, they would be transformed into serpentinites composing mainly of serpentine and magnetite, and this process called is "serpentinization".
The transparent or semi-transparent serpentines minerals, together with the black opaque magnetite, give rise to Taiwan black jade.

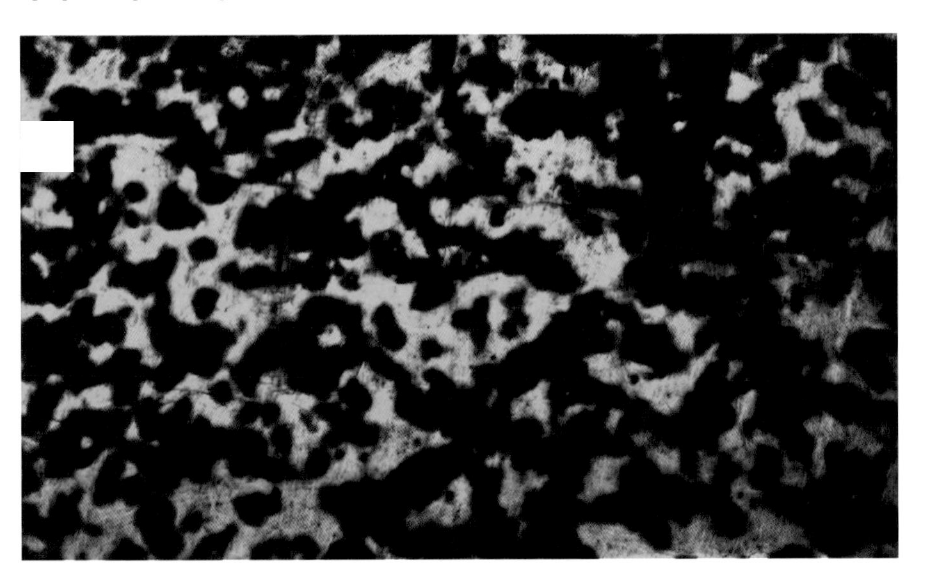

玉石小博士 - 常見玉石的鑑別
Greeting from Dr. Jade: Identifying Common Jade Stones

臺灣墨玉與其他世界常見的玉石 - 岫岩玉、閃玉、輝玉，到底有何相似或差異之處呢？讓我們一起來瞧一瞧：
A look at the table below helps us understand the differences and similarities between Taiwan black jade and other common jade stones in the world, notably Xiuyan jade, nephrite (Taiwan jade), and jadeite .

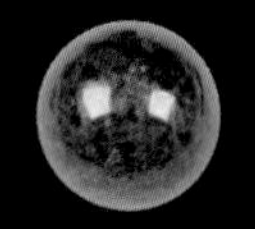
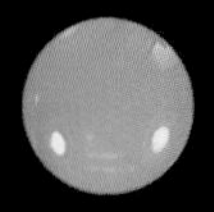
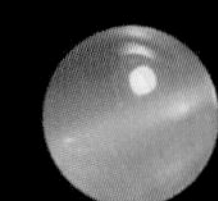

	臺灣墨玉	岫岩玉	閃玉(臺灣玉)	輝玉
礦物組成	蛇紋石	蛇紋石	透閃石-陽起石	含鈉的輝石
化學成分	$(Fe,Mg)_3Si_2O_5(OH)_4$	$(Fe,Mg)_3Si_2O_5(OH)_4$	$Ca_2Mg_5Si_8O_{22}(OH)_2$ $Ca_2Fe_5Si_8O_{22}(OH)_2$	$NaAlSi_2O_6$
硬度	4.5	4.5-5.5	5.5-7	6.5-7
比重	2.4-2.8	2.6-2.8	2.9-3.1	3.5-3.6
折射率	1.56-1.70	1.55-1.56	1.60-1.63	1.65

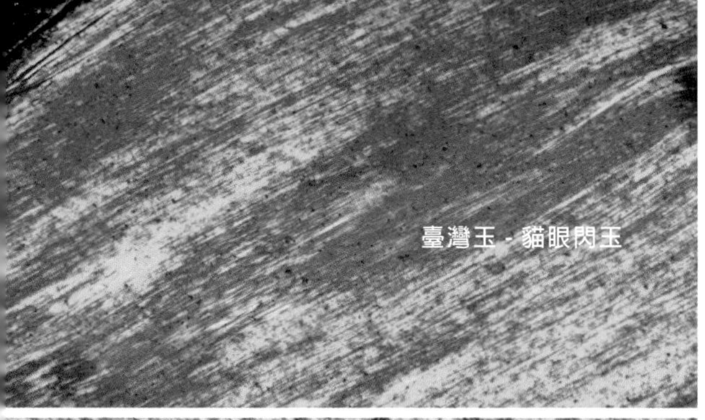

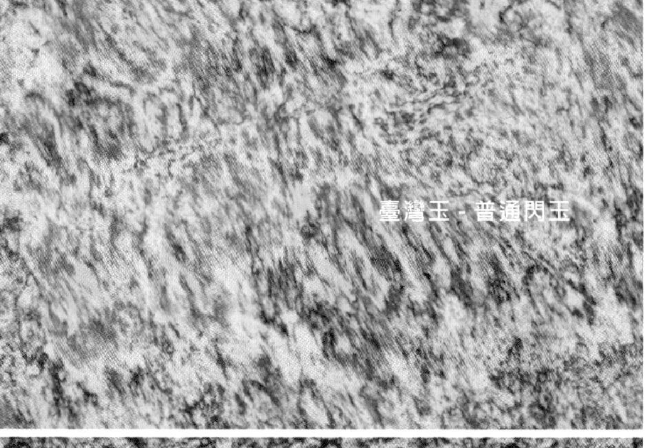

玉石萬花筒

Jade Stone through Kaleidoscope

針對寶石礦物的研究，常將礦物磨成 0.03mm 厚的透光薄片，利用偏光顯微鏡來觀察，透過光線穿透不同礦物的顏色變化，寶石學家得以瞭解礦物的種類、數量比例及岩理組織。

In gemological studies, minerals are usually ground into translucent thin sections with a thickness of approximately 0.03 mm, which in turn can be observed using a polarizing microscope. Different colors are generated as light penetrates through different minerals, thereby helping gemologists identify the types, amounts, ratios, and textures of the minerals.

臺灣玉 - 貓眼閃玉 (Cat's-eye Nephrite)：由長度超過 1000 微米纖維狀的透閃石礦物結晶組成，具同一方向排列，如果垂直纖維結晶琢磨，可呈現貓眼現象。

Taiwan Jade – Cat's-Eye Nephrite: This special nephrite is composed of the fibrous, unidirectional crystals (with a length exceeding 1000 μm) of tremolite. Grinding and polishing the fibrous crystals vertically leads to the emergence of the shape "cat's-eye."

臺灣玉 - 普通閃玉 (Common Nephrite)：由 40 至 150 微米纖維狀透閃石礦物結晶組成。

Taiwan Jade – Common Nephrite: Like it's cats-eye counterpart, this regular nephrite is also composed of the fibrous crystals of tremolite; the length of the crystals, however, ranges between 40 to 150 μm.

臺灣墨玉：主要由葉蛇紋石礦物組成，結晶比蛇紋岩中者細緻均勻，黑色不透明處為磁鐵礦。

Taiwan Black Jade: Mainly composed of crystals similar to those of antigorite (a polymorph of serptentine); its crystals are finer and more even than those of serpentine. The black non-transparent parts are magnetite.

蛇紋岩：主要由葉蛇紋石礦物組成，結晶比臺灣墨玉者粗，也常含如直閃石特別大結晶的礦物等，結晶黑色不透明處為磁鐵礦。

Serpentinite: Mainly composed of crystals similar to those of antigorite; its crystals are coarser than those of Taiwan black jade. Minerals with larger crystals, such as anthophyllite, are often found in serpentinite. The black non-transparent parts are magnetite.

世界重要玉石分布地

Major Deposits of Jade Stones around the World

美麗的寶石人見人愛，你知道世界上重要的輝玉、閃玉及蛇紋石玉等玉石，產在哪裡嗎？讓我們按圖索驥 - 在地圖找找看囉。
不要忘記，我們臺灣就出產閃玉（臺灣玉）及蛇紋石玉（臺灣墨玉）等兩種玉石。

Beautiful gemstones are everyone's darlings! However, do you know where the major deposits rich in nephrite, jadeite, and serpentine jade around the world are? Get ready to put them on the map!
Don't forget that, here in Taiwan, we have both nephrite (Taiwan jade) and serpentine (Taiwan black jade).

閃玉 (藍色標記)

Nephrite (marked in blue)

1.新疆 崑崙山 2.遼寧 寬甸 3.四川 汶川 4.江蘇 溧陽
5.新疆 天山 馬納斯 6.河南 淅川 7.臺灣 花蓮 8. 俄羅斯聯邦的圖瓦共和國 薩彥嶺區 9.大韓民國 江原道 春川市
10.巴布亞紐幾內亞 新幾內亞東南部 11.新喀里多尼亞 洛易提群島 12.澳洲 科威爾 13.澳洲 新南威爾斯 坦瓦斯
14.紐西蘭 南島 15.加拿大 英屬哥倫比亞 托拿根河 16 加拿大 英屬哥倫比亞 夫拉則河 17 美國 阿拉斯加州 科伯克河
18.美國 懷俄明州 蘭德 19.美國 加州 門德奇諾郡 20.美國 加州 瑪瑞波薩郡 21.美國 加州 蒙特瑞郡 22.波蘭 奧得河
23.瑞士 格勞賓頓 24.義大利 里維耶拉 25.辛巴威 西呂科威東南區

1. Kunlun Mountains, Xinjiang Province, China
2. Kuandian, Liaoning Province, China
3. Wenchuan, Sichuan Province, China
4. Liyang, Jiangsu Province, China
5. Manasi, Tien Shan, Xinjiang Province, China
6. Zhe Chuan, Henan Province, China
7. Hualien, Taiwan
8. Sayan Mountains, the Republic of Tuva
9. Chuncheon City, Gangwon Province, Korea
10. Southeastern Papua New Guinea
11. Loyalty Islands, New Caledonia
12. Cowell, Australia
13. Tamworth, New South Wales, Australia
14. South Island, New Zealand
15. Turnagain River, British Columbia, Canada.
16. Fraser River, British Columbia, Canada
17. Kobuk River, Alaska, the United States
18. Lander, Wyoming, the United States
19. Mendocino County, California, the United States
20. Mariposa County, California, the United States
21. Monterey County, California, the United States
22. Oder River, Poland
23. Graubunden, Switzerland
24. Riviera, Italy
25. Southeast corner of Zimbabwe

輝玉 (橘色標記)

Jadeite (marked in orange)

1.緬甸 克欽邦 2.日本 新潟縣 3.瓜地馬拉 蒙太格河 4.美國 加州 聖彼納多郡

1. State of Kachin, Myanmar
2. Nigata Perfecture, Japan
3. Motagua River, Guatemala
4. San Benito County, California, the United States

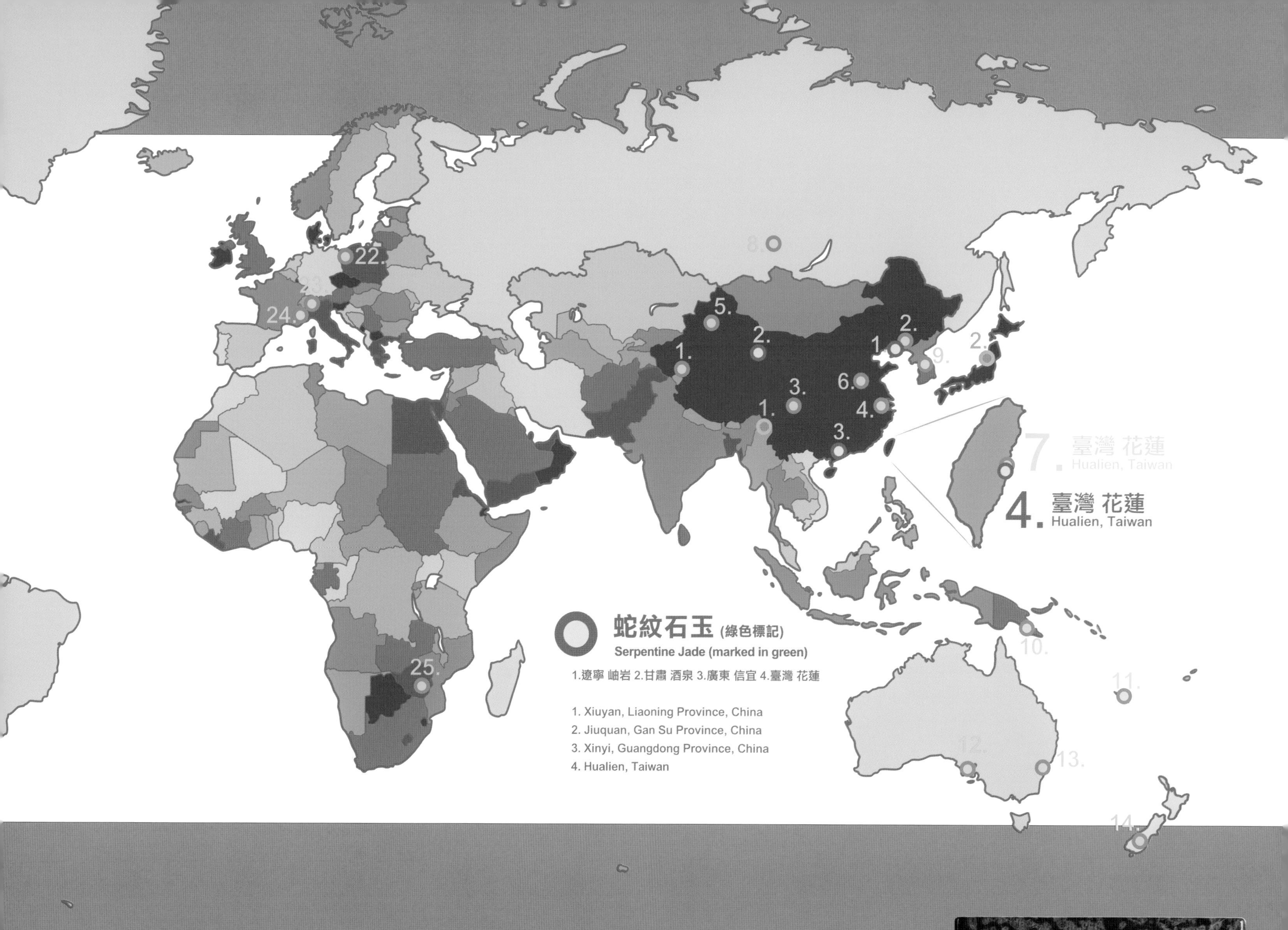

8.
22.
23.
24.
5.
2.
1.
3.
1.
6.
4.
3.
2.
1.
9.
2.
7. 臺灣 花蓮
Hualien, Taiwan
4. 臺灣 花蓮
Hualien, Taiwan
蛇紋石玉 (綠色標記)
Serpentine Jade (marked in green)
1.遼寧 岫岩 2.甘肅 酒泉 3.廣東 信宜 4.臺灣 花蓮
1. Xiuyan, Liaoning Province, China
2. Jiuquan, Gan Su Province, China
3. Xinyi, Guangdong Province, China
4. Hualien, Taiwan
25.
10.
11.
12.
13.
14.

臺灣玉與臺灣墨玉的

化學成分

Taiwan Jade vs. Taiwan Black Jade: Chemical Composition

臺灣玉與臺灣墨玉雖然共生在一起，但卻是由不同礦物種類組成，因此化學成分也有差異。臺灣玉是透閃石（$Ca_2Mg_5Si_8O_{22}(OH)_2$）與陽起石（$Ca_2Fe_5Si_8O_{22}(OH)_2$）的固溶體，常含有黑色的鉻鐵礦（$FeCr_2O_4$）；臺灣墨玉則是蛇紋石（$Mg_3Si_2O_5(OH)_4$）含有黑色的磁鐵礦（Fe_3O_4）。藉由國立臺灣博物館所有的精密儀器 -X 光螢光分析儀（XRF），對臺灣玉與臺灣墨玉進行掃描，可以清楚看清兩者化學成分的不同。

Though in association, Taiwan Jade and Taiwan black jade are composed of different minerals and manifest different chemical compositions.Taiwan jade is a solid solution of tremolite [$Ca_2Mg_5Si_8O_{22}(OH)_2$] and actinolite [$Ca_2Fe_5Si_8O_{22}(OH)_2$], usually dotted with black chromite ($FeCr_2O_4$). Taiwan black jade, on the other hand, is serpentine [$Mg_3Si_2O_5(OH)_4$] disseminated with black magnetite (Fe_3O_4).
One can observe clearly the chemical difference of Taiwan jade and Taiwan black jade, using the XRF (X-ray fluorescence) analyzer at National Taiwan Musuem.

利用 X 光螢光分析儀 (XRF) 掃描玉石樣品，可以瞭解其主要組成化學成分。臺灣玉由矽、鎂、鈣、鐵元素組成，其黑色共生礦物 - 鉻鐵礦由鐵及鉻元素組成；臺灣墨玉由矽、鎂元素組成，其黑色共生礦物 - 磁鐵礦由鐵元素組成。
Using an X-ray fluorescence analyzer to scan samples of jade stone helps us understand its chemical composition. The major elements of Taiwan jade include silicon, magnesium, calcium, and iron, and the black mineral association is chromite composed of iron and chrome. For Taiwan black jade, the major elements are silicon and magnesium with the black mineral association of magnetite composed of iron.

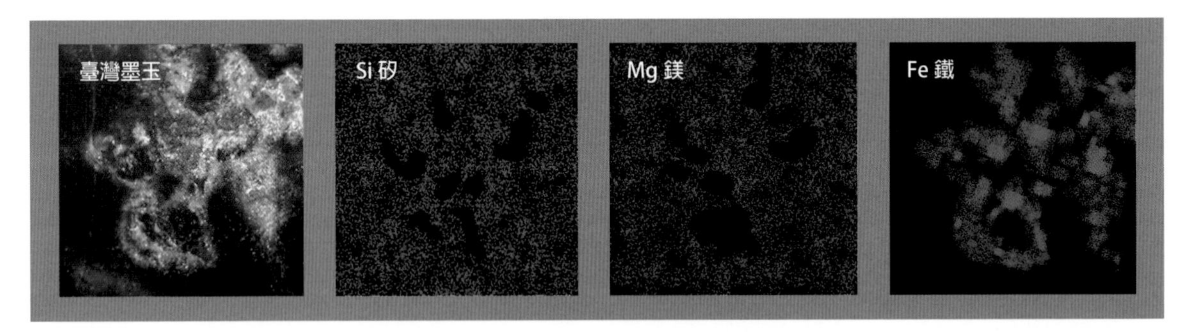

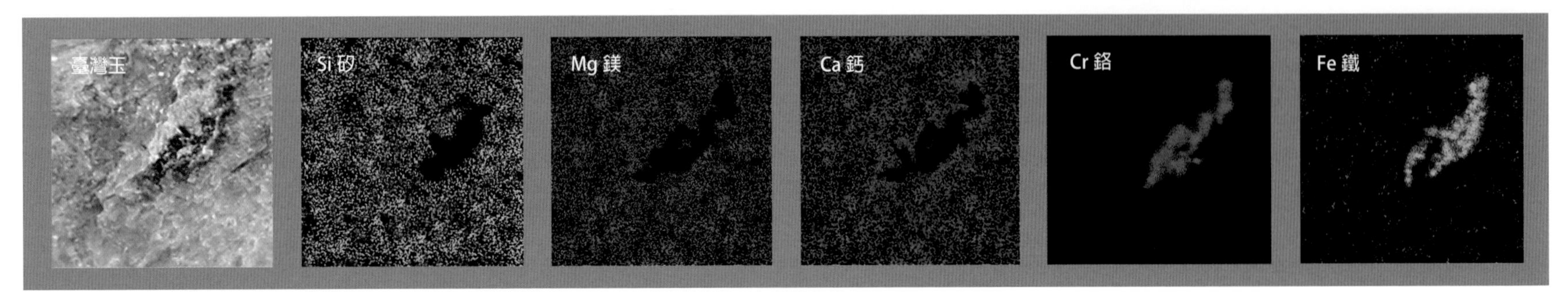

第四章：玉不琢不成器，玉石的加工

Chapter 4 : Processing of Jade

墨玉生活用品 - 茶具組

墨玉生活用品 - 筷與擺

玉石的加工

Processing of Jade

玉石在露天或坑道中，經開採後運送下山，大塊玉料依質地分類，經適當切割取材後，依設計需求進行加工。一般玉的加工分為切割 - 去瑕 - 磨型 - 拋光等流程，才能成為有用器物。

Raw jade stones are exploited, either in the open or in a tunnel, and transported to a processing plant where they are categorized based on texture and then subjected to proper processing in line with the desired designs.Raw jade stone needs to be processed into a useful artifact through proper and effective processing that incorporates defect-removing, cutting, grinding, shaping, polishing, refining and related steps.

工序一：玉石切割
Step 1:Cutting

切割取材，去邊料與裂紋。

Have the desired portion of the jade stone cut; scraps and fissures are then trimmed.

工序二：研磨去瑕
Step 2:Grinding for flaw removal

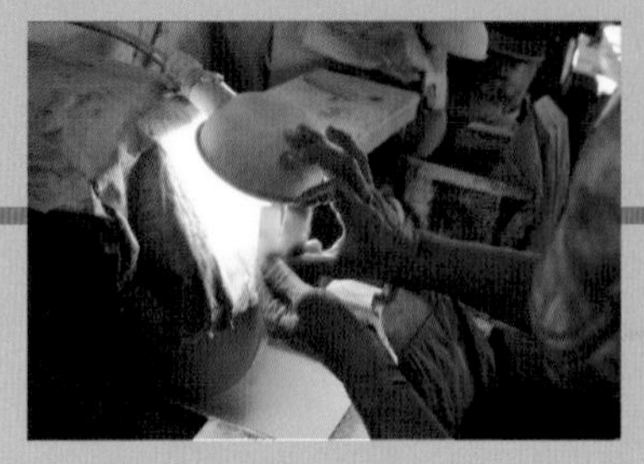

表面黑點瑕疵磨掉，燈光下可見其玉質剔透。

Grind the surface to get rid of flaws and defects like tiny black smears. As shown in the photograph, the beautiful translucence of the jade stone reveals itself under the glittering light.

工序三：造形修整
Step 3:Trimming and Shaping

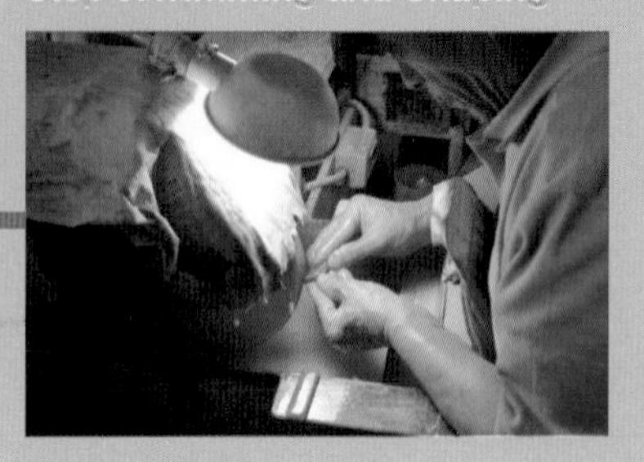

造型研磨至所要求的形狀。

Continue the grinding and trimming until the desired shape starts to emerge.

工序四：表面拋光
Step 4:Surface polishing

砂紙拋光依序使用400#、800#、1200#等。

Use sandpaper, respectively in the order of 400#, 800#, and 1200#, to polish the surface.

工序五：皮革輪拋光
Step 5: Further Polishing

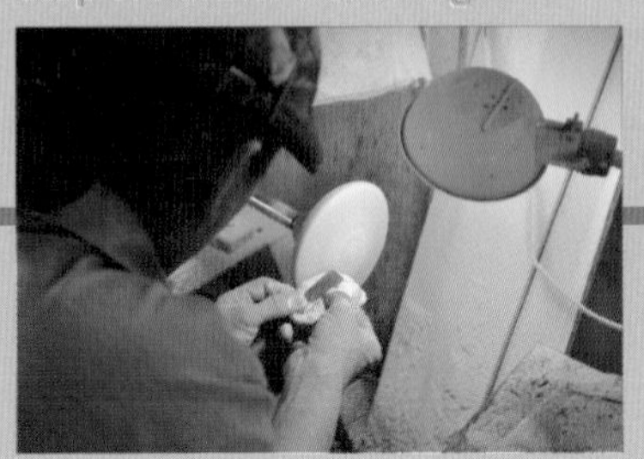

皮革輪加拋光粉做最後一道程序的拋光。

Use grinding wheel with burnishing powder to complete further and final polishing.

工序六：鑽孔
Step 6:Hole grinding

利用超音波震動混加金鋼砂鑽孔。

Drill the hole with ultrasonic vibration technique plus carborundum.

工序七：組合金工首飾
Step 7:Final assembly and embellishment

玉石組銀墜子頭，成為可配戴的首飾品。

Attach the desired metalwork or jewelry to the processed jade stone like the jade pendant with a silver ear shown in the photograph.

第五章：臺灣玉石精品

Chapter 5 : Taiwan Jade Boutique

傳統巧雕
Exquisite

國人對玉石的情感甚篤，相信玉石具辟邪護身與增福添壽等功能，所以玉石的雕刻工藝自古即很發達，大至山水意境，祥龍瑞鳳，小至擺件把玩，首飾巧雕，相當討喜可愛。
在玉石雕刻創作題材上，常以福祿壽喜寓意祥瑞，梅蘭竹菊等具君子雅士之風尚為主要內容。

Chinese have long demonstrated a particular fondness of jade artifacts which are believed to be capable of warding off evil spirits, safeguarding physical and spiritual wellbeing, increasing good luck, and inviting longevity. Jade carving has accordingly been practiced and celebrated as an art form since ancient times; jade artifacts come in all sizes with a wide spectrum of enchanting images, ranging from jade screens carved with expansive landscape and an auspicious pair of dragon and phoenix to jade pendants marked with an exquisite twin-fish pattern and a majestic peony in full bloom, symbolizing harmony, beauty, and prosperity.
In addition to the traditional "four treasures" (fortune, prosperity, longevity, and happiness), the "four gentlemen among flowers" – plum, orchid, bamboo, and chrysanthemum – and other images embodying admirable virtues are also major themes in jade carving.

墨玉礁魚

臺灣玉金玉佩飾

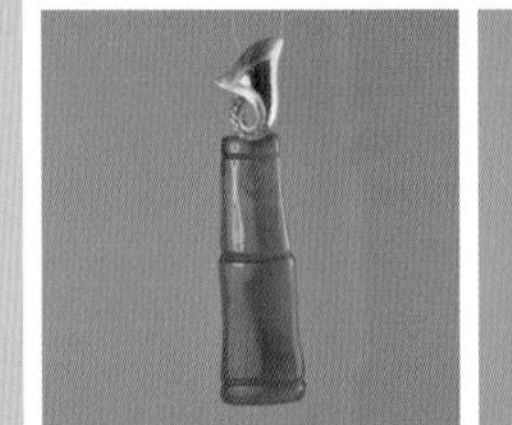
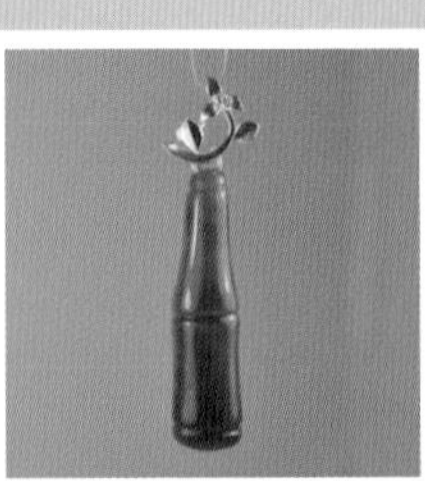

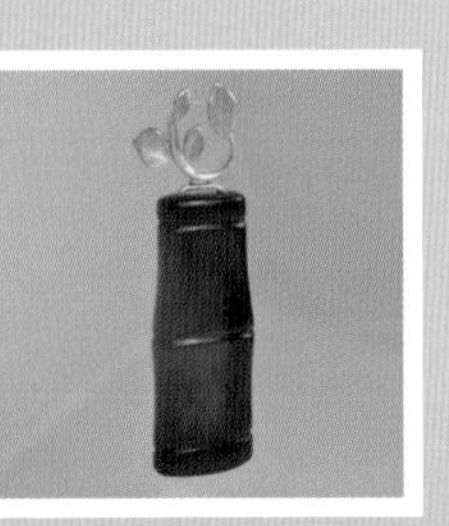

以臺灣玉為材料所製作的首飾雕件

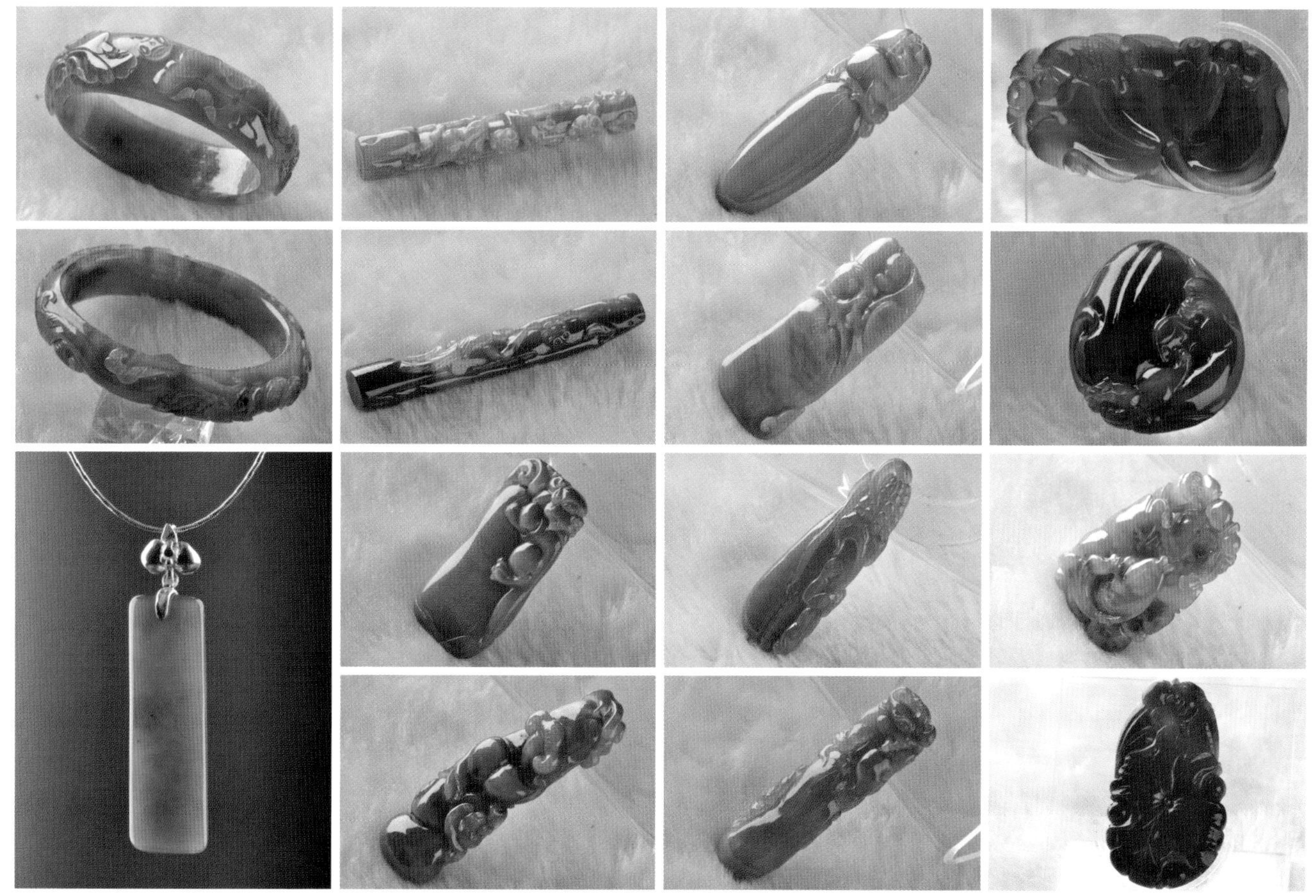

以臺灣玉為材料所製作的傳統雕件

01. 荷塘戲波
02. 喜鵲登梅
03. 玉石相倚春意鬧
04. 富貴在天
05. 花開富貴
06. 雙鯉啓福
07. 魚荷雙喜

以臺灣玉為材料所製作的傳統雕件

01. 祥獸獻瑞
02. 螳螂與翠玉白菜
03. 喜鵲登枝
04. 龍鯉吐珠
05. 鳥語花香

以臺灣墨玉為材料所製作的傳統雕件

01. 臺灣牛
02. 有餘
03. 虎威
04. 獅吼
05. 海中鮫

01

02

03

04

05

以臺灣墨玉為材料所製作的傳統雕件

01. 龍如意
02. 橫行霸道
03. 雞鳴報曉
04. 雙蝠抱李
05. 駿馬

02

04

01

03

05

宗教神學

Theological and metaphysical significances of Jade

在宗教神學玄學中，以為天地萬物皆有靈性，而經千萬年時間孕育生成之翠玉寶石，因其稀有且剔透質韌，常被利用為朝供天地與祭祀活動用的禮器。
天體運行，日、月、星、辰，各有軌跡，宇宙秩序輾轉不停，此皆以北極星及各個天地方位為軸。東方民族，常以方位，例如東、西、南、北，十天干，十二地支，為眾神之根源。自遠古時代，東方人對所有神祇都抱持敬畏的態度。玉器常被古人用來做為儀典的祭器，或當做供奉神明的祭禮，以表達他們對神祇的崇敬態度。

In Chinese metaphysics, all things in heaven and on earth are believed to carry with them unique spirits. Having existed for thousands of years, far longer than any human beings and dynasties, gemstones like jade and emerald are gifts from gods and thus often processed into ceremonial and sacrificial vessels.
The revolution of the celestial body, the Sun, the moon, and the stars, follows its own path, and the sequence of the universe tosses and turns, which all center on Polaris and various positions.For Asians, positions, such as the East, the West, the South, the North, the ten Heavenly stems, and the twelve Earthly Branches, are the origins of different Gods. Since the remote antiquity, Asians hold in awe to all sorts of Gods. The Jade articles serve as ritual vessels or sacrificial offerings to represent the ancients' profound reverence to Gods.

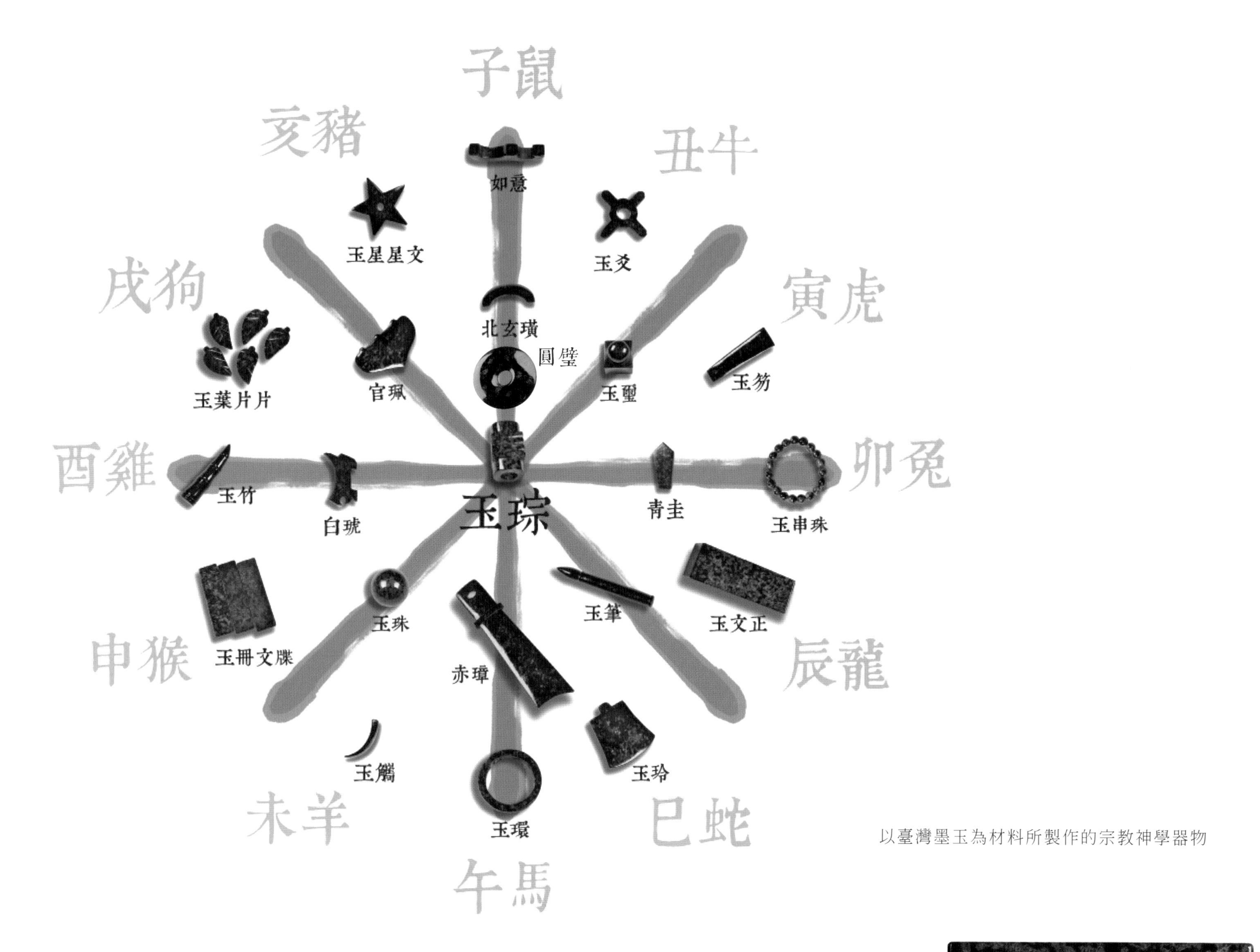

以臺灣墨玉為材料所製作的宗教神學器物

臺灣墨玉宗教神學器物

時尚新設計

Innovative New Design

在時空背景的轉換下，傳統的玉石首飾，融入當代文化新思維與視覺新元素，可發展成為具時尚感的首飾設計及文創工藝。
本區以臺灣玉的故鄉—花蓮為創意萌發的起點，由國立東華大學藝術創意產業學系創作有關花東地域人文環境與歷史發展的"洄瀾飾"系列時尚新設計。

As time changes and culture evolves, the design and development of traditional jade artifacts are beginning to experience a renaissance in the wake of cultural and creative industries that introduce new aesthetic elements and innovative artistic insights.
In this section, we present to you the "Huilan Ornament" series of fashionable and innovative jade artifacts designed by the faculty members and students from the Department of Arts and Creative Industries at National Dong Hwa University, who are privileged to have their creativity inspired by the historical, cultural, and humanistic heritage of Hualien, the home of Taiwan jade.

迷路森林

當所倚賴的生存空間成為殺戮戰場，大自然的因果循環，終將回饋到人類身上。謹記經驗教訓，人類開始學習生態保育的重要與對生存權的相互尊重。

綠野仙蹤

以鹿野的生態歷程為發想。復育，重現風華，讓綠野再現仙蹤。

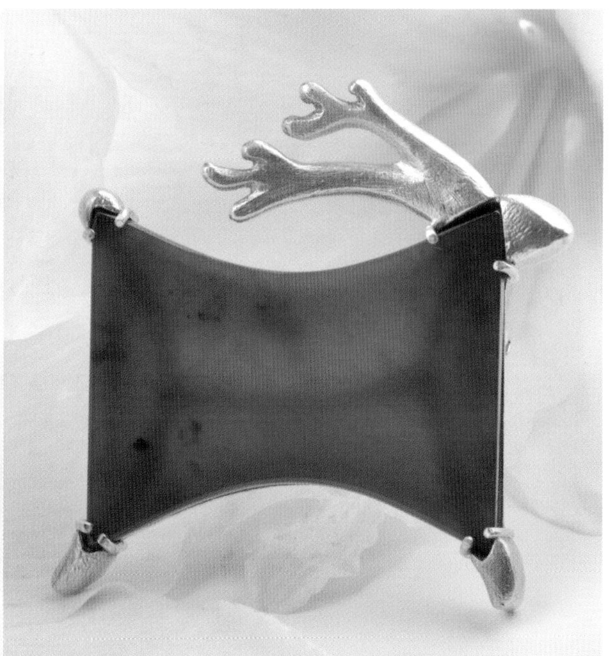

東華樂園 I

以東華三寶，環頸雉，古早兔與清明草，結合金線蛙與山豬，自然環境與人文教育的交集，花蓮 - 東華，真是幸福的樂園。

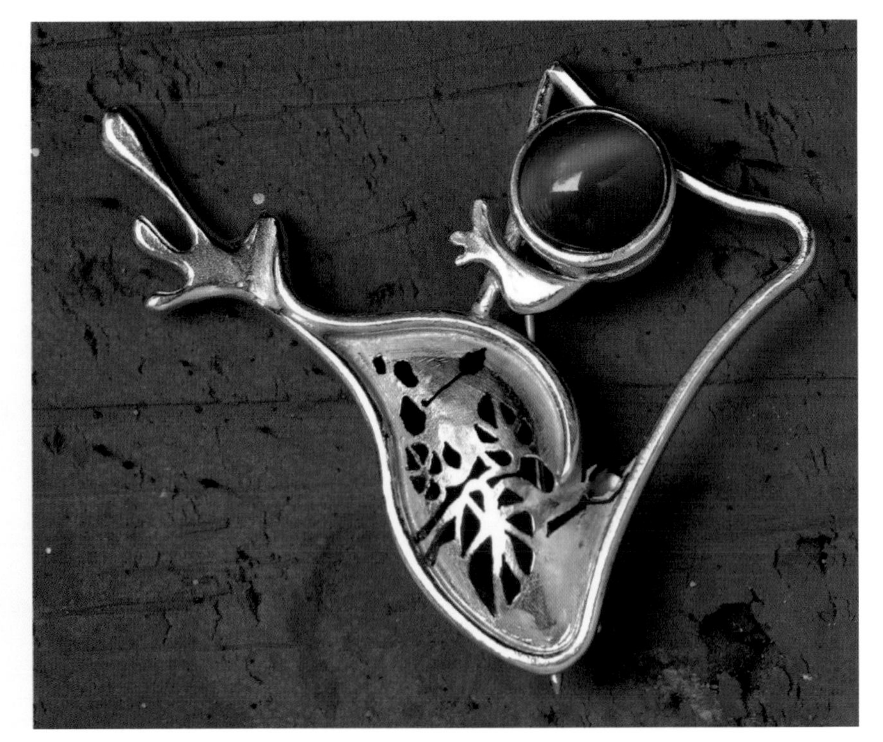

東華樂園 II

快樂天堂

東華校園祥和而充滿愛心，流浪動物得以在此域安養無虞。

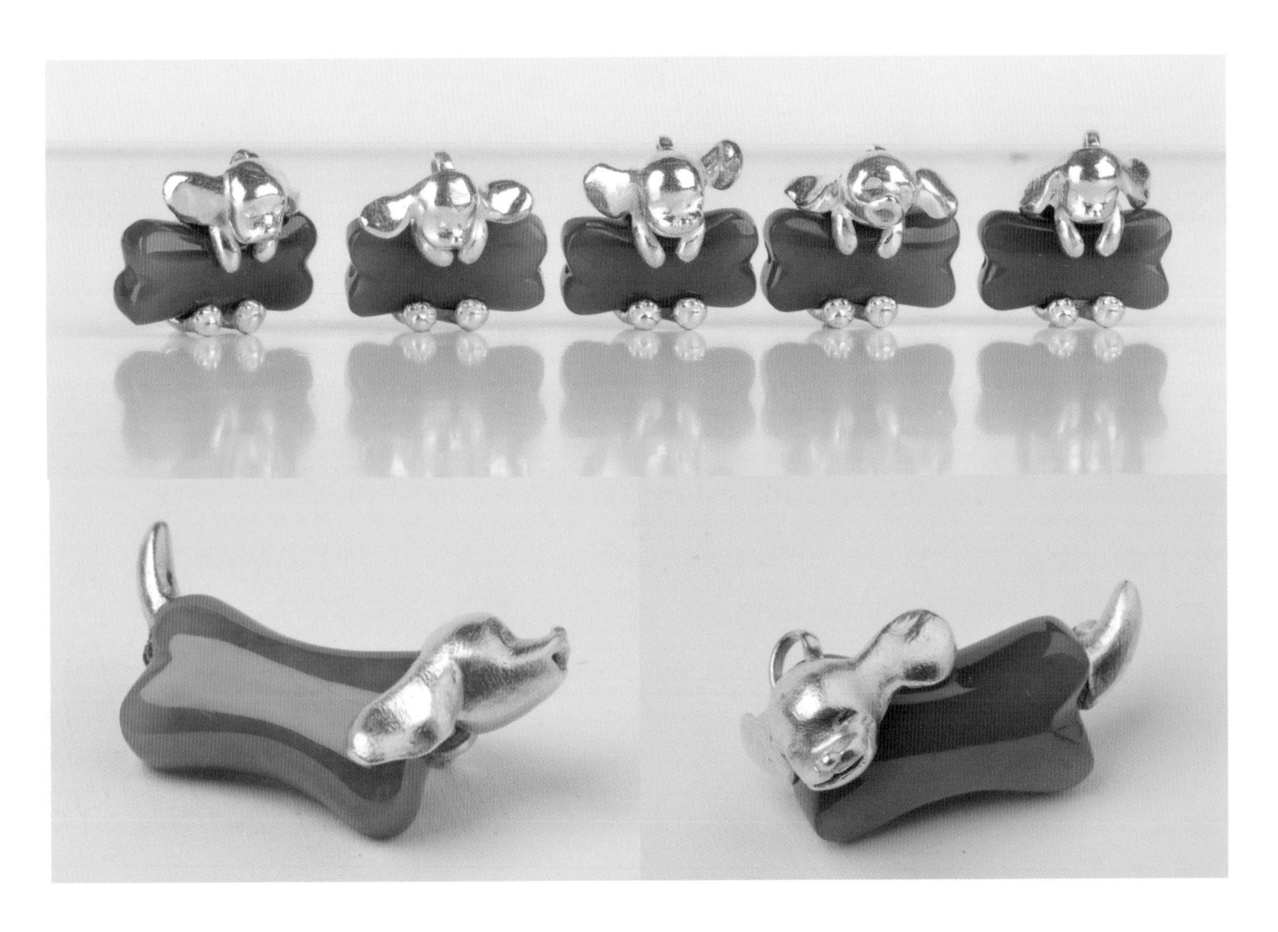

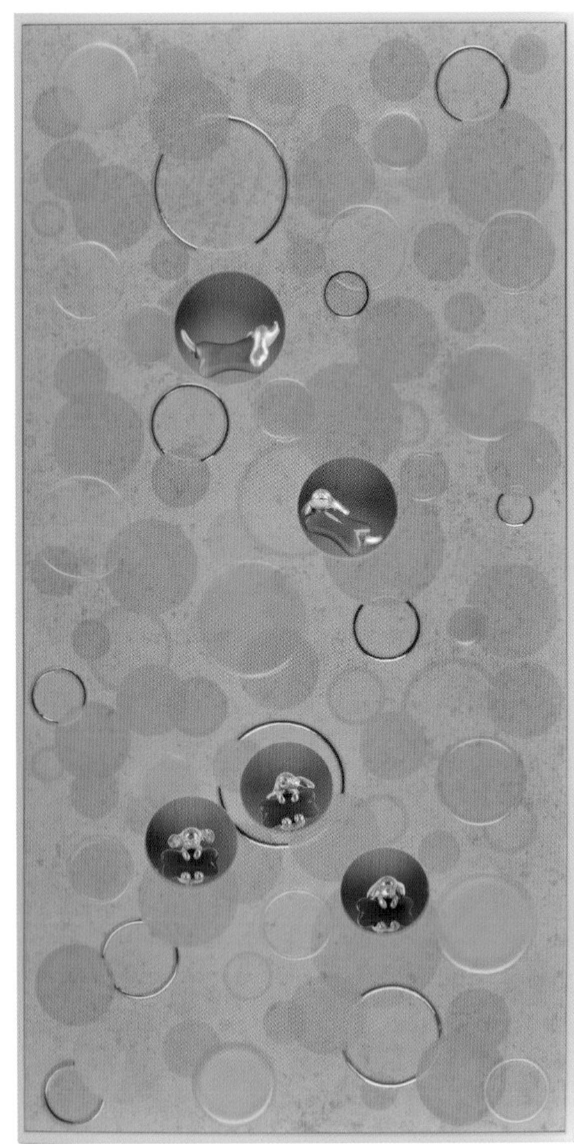

日照後山

花東地域擁有得天獨厚的天然資源，給予合適的發展機會，
將會如傑克的豌豆般，發展出令人刮目相看的神奇結果。

後花園

後花園的美，三分天然，七分想像。擷取美的節點串連，讓畫面成為翩翩飛起的感動，結實纍纍的收穫，讓觀賞者心嚮往之。

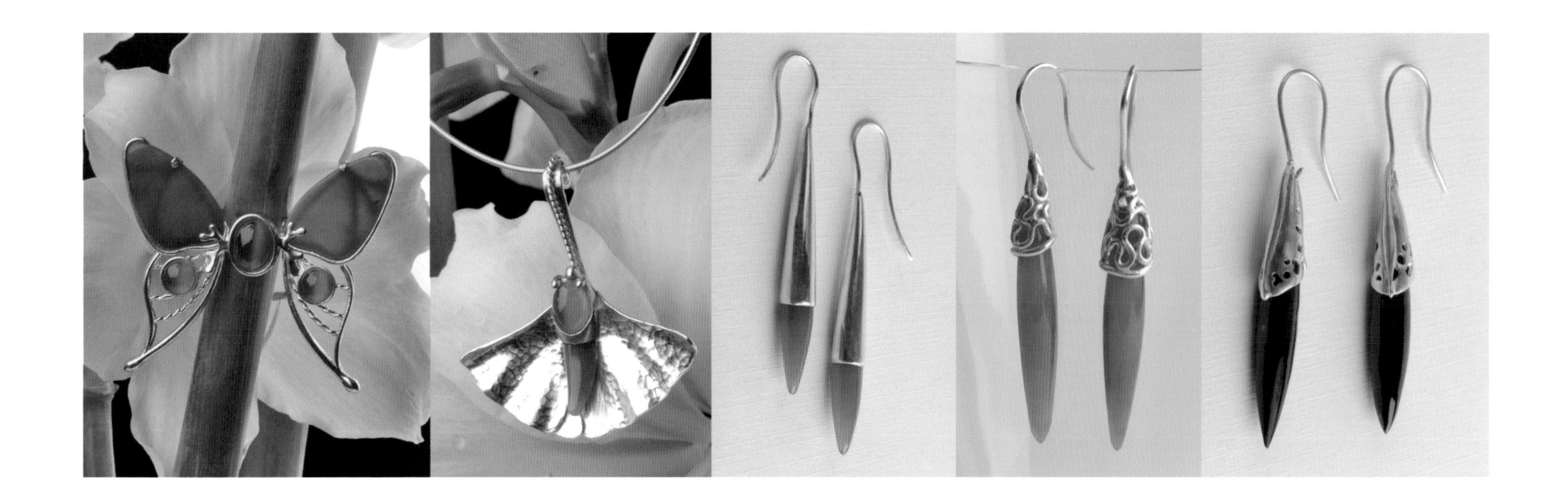

種豆得豆

種福豆，得福豆。種子賦予可達成願望的神奇使命，用心植栽，所收成的結果，必定令人刮目相看。

希望種子

「要怎麼收穫，先那麼栽」。
期待「愛情」嗎？埋顆愛的種子吧！待發芽後，會心花朵朵開。所以，「幸福」、「財富」與「幸運」，你想選擇哪一種？

生機

眼，神也！貓眼玉的閃耀生動，如慧詰的靈魂之窗，是生命與生態的交織樂章，讓觀賞者心嚮往之。

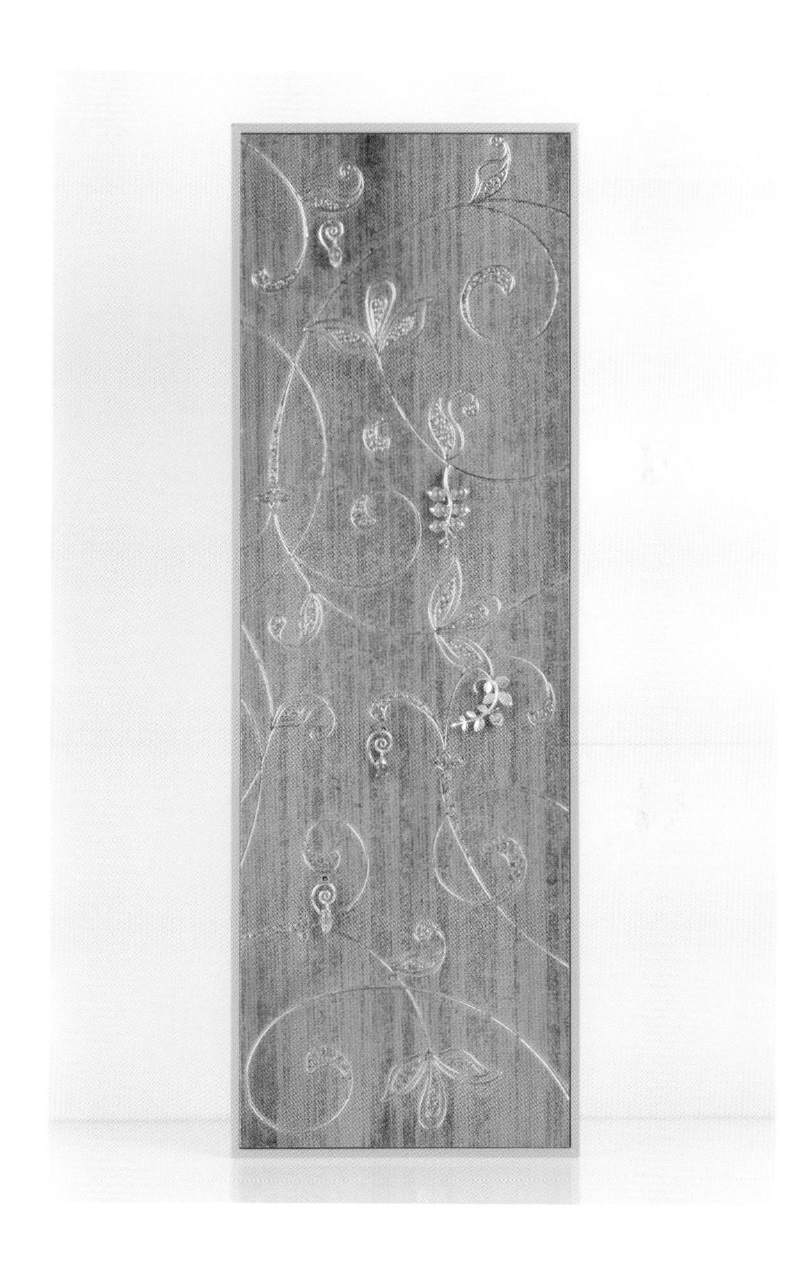

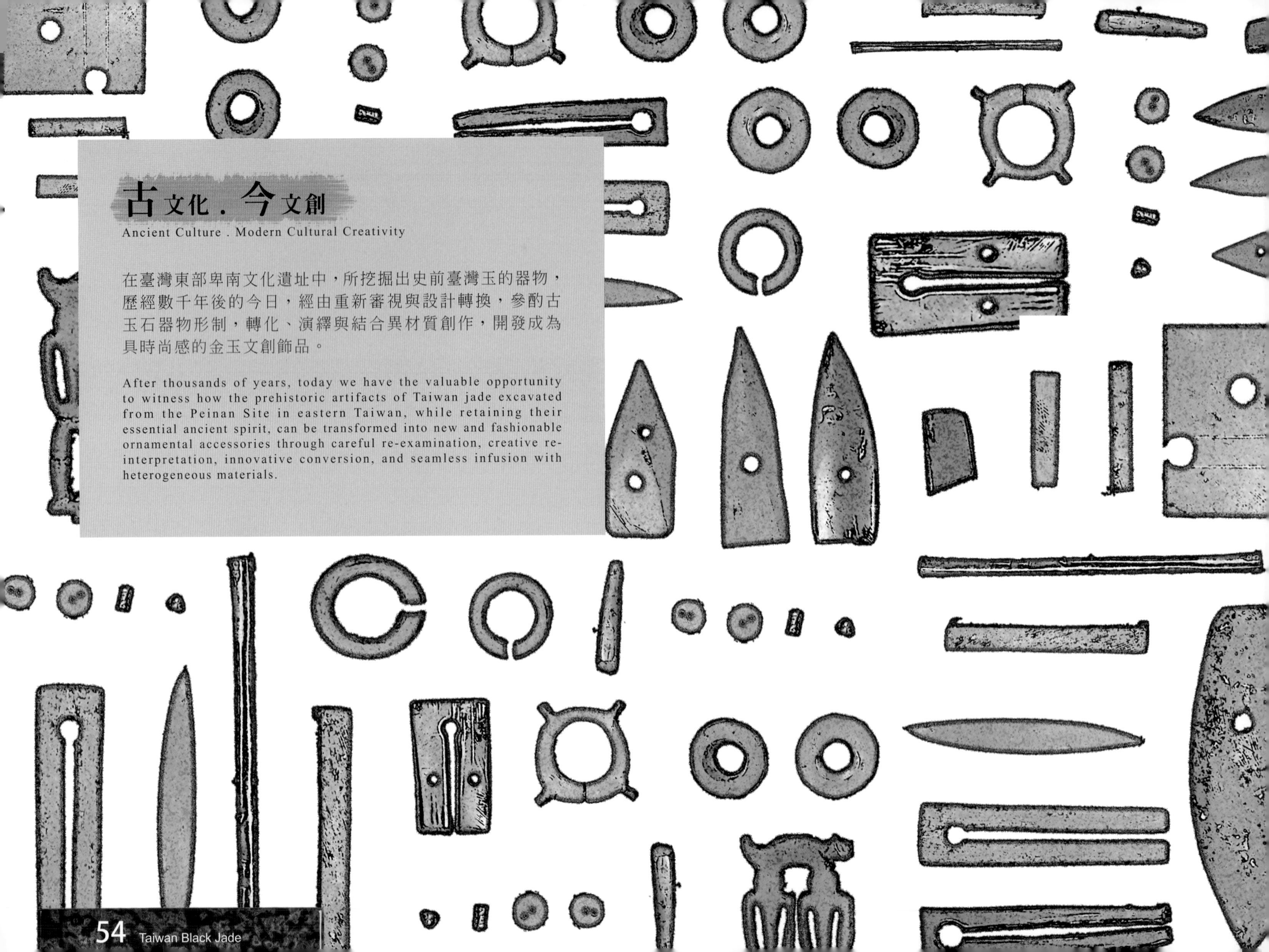

古文化．今文創

Ancient Culture . Modern Cultural Creativity

在臺灣東部卑南文化遺址中，所挖掘出史前臺灣玉的器物，歷經數千年後的今日，經由重新審視與設計轉換，參酌古玉石器物形制，轉化、演繹與結合異材質創作，開發成為具時尚感的金玉文創飾品。

After thousands of years, today we have the valuable opportunity to witness how the prehistoric artifacts of Taiwan jade excavated from the Peinan Site in eastern Taiwan, while retaining their essential ancient spirit, can be transformed into new and fashionable ornamental accessories through careful re-examination, creative re-interpretation, innovative conversion, and seamless infusion with heterogeneous materials.

以臺灣玉及臺灣墨玉為材料所製作的新文創飾品

卑南仿古墜飾，墜鍊，頸飾，胸飾

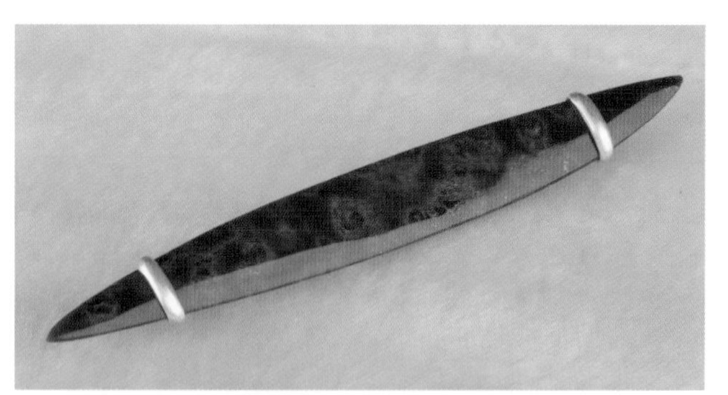

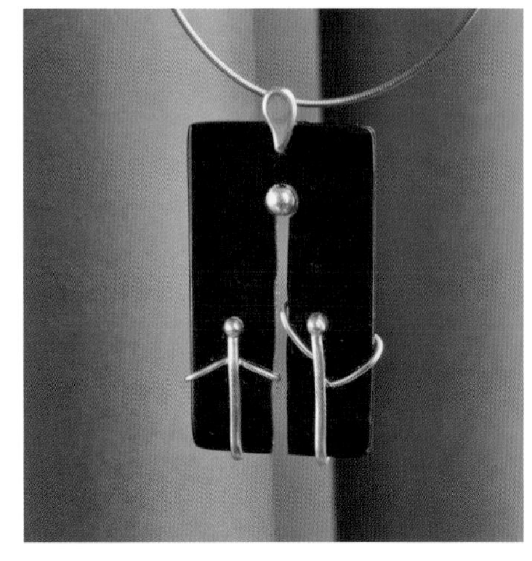

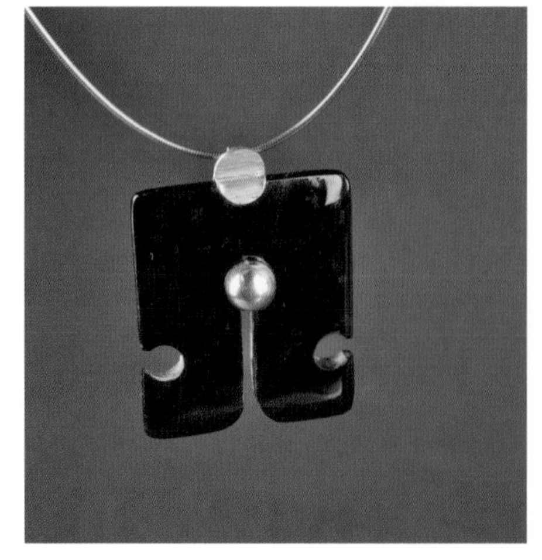

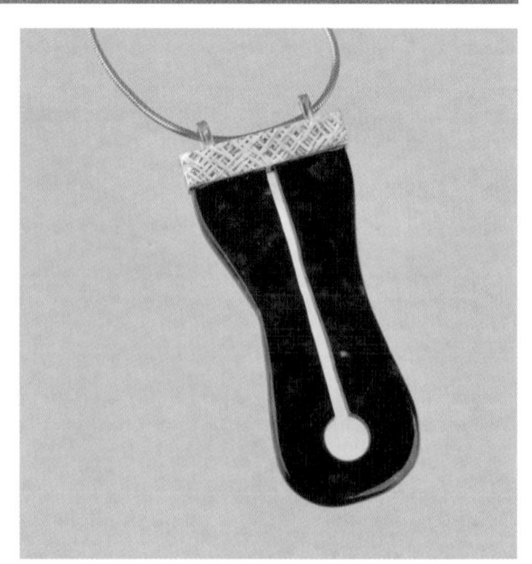

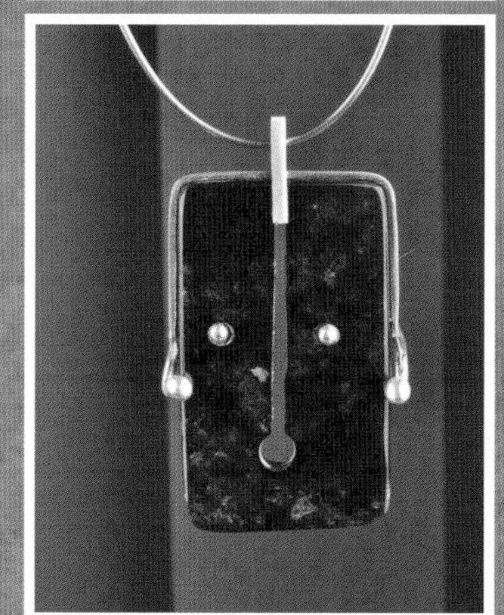

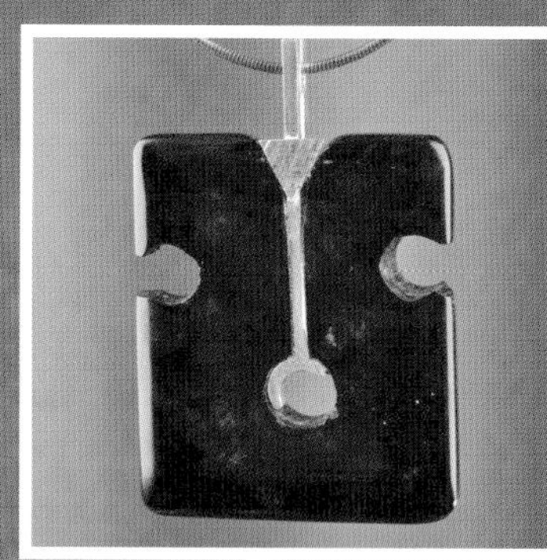

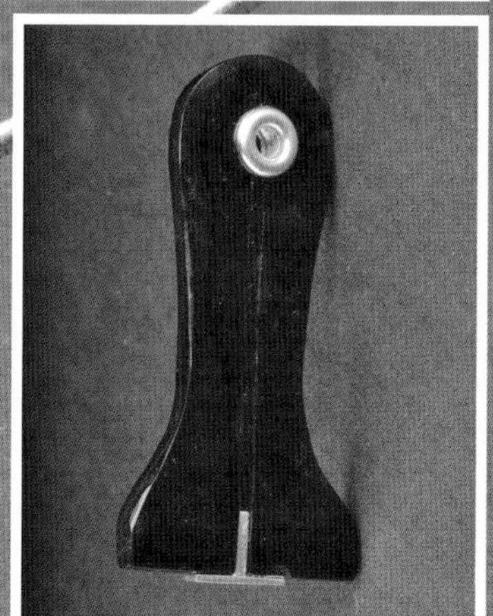

臺灣墨玉卑南文創飾品

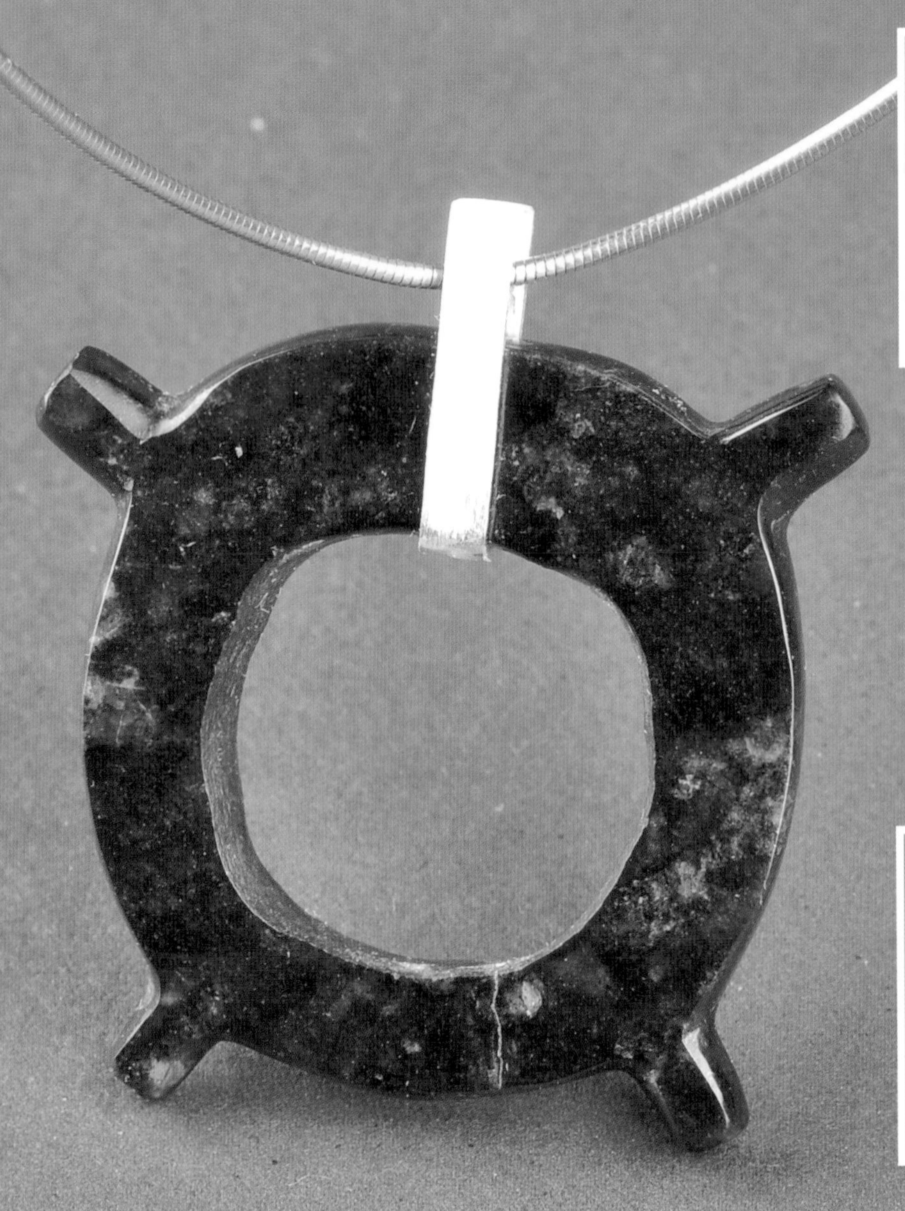

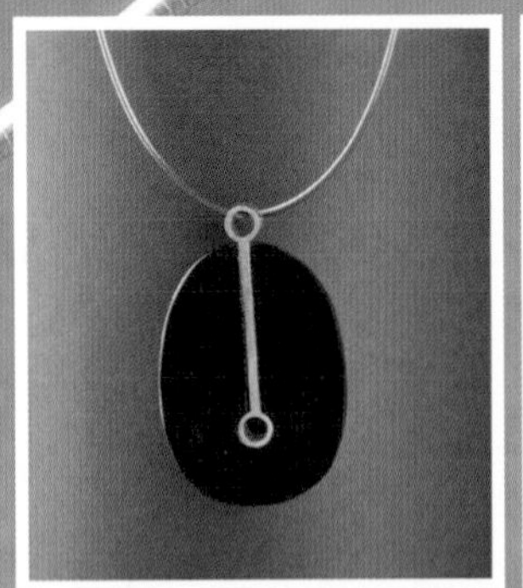

臺灣玉與臺灣墨玉 卑南仿古墜鍊

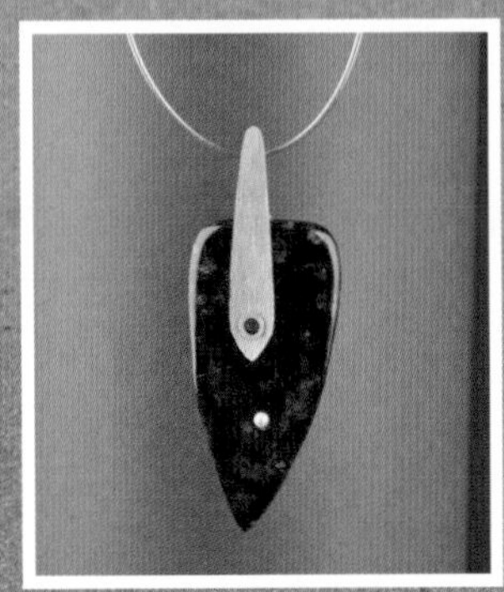

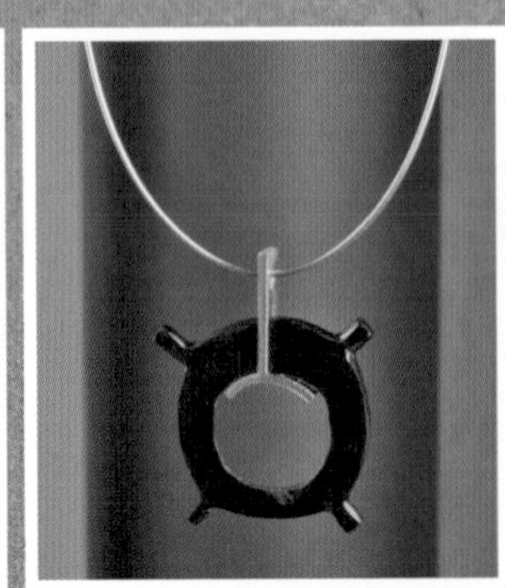

臺灣玉與臺灣墨玉
卑南仿古飾品

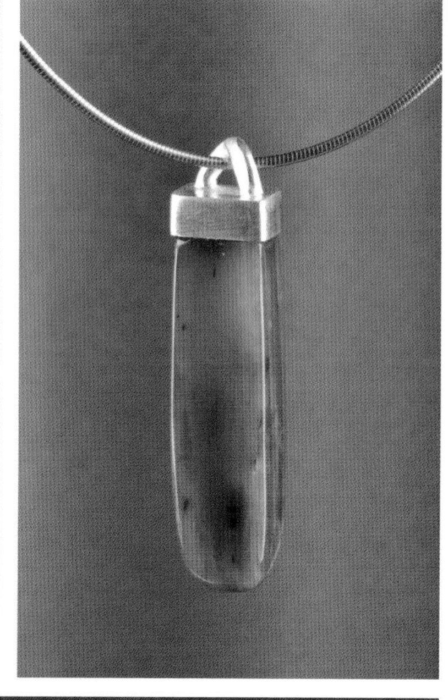

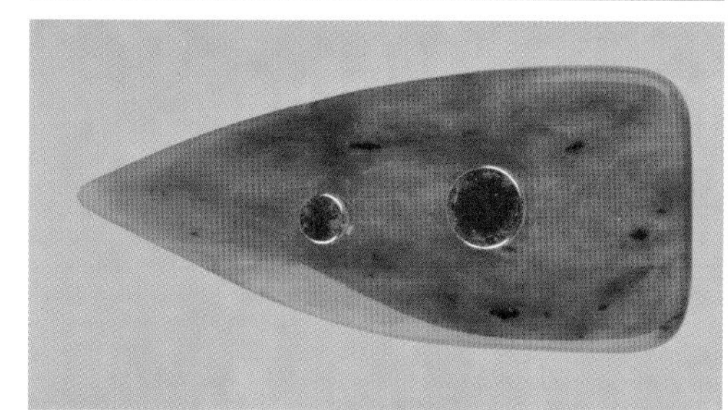

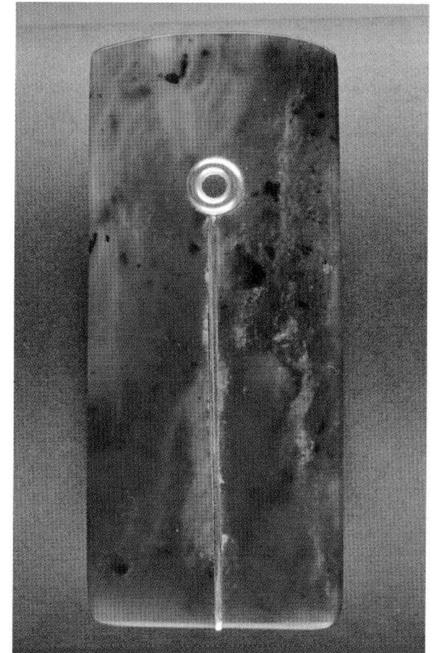

發行人：洪世佑
出版委員：洪世佑 王逸群 許毓純 方建能 吳嘉琦
黃星達 彭騰衝 劉美珠 方瓊德
作者：方建能 林淑雅
助理編輯：李知矩
美術設計：睛豔國際有限公司 蔡力強
出版者：國立臺灣博物館
地址：10046 臺北市襄陽路 2 號
電話：(02)2382-2699
網址：http://www.ntm.gov.tw
電子郵件：jtmeditor@ntm.gov.tw
出版日期：中華民國 103 年 3 月（初版）
中華民國 108 年 10 月（三刷）
定價：新臺幣 220 元
展售處：
國立臺灣博物館員工消費合作社 10046 臺北市襄陽路 2 號 (02)2371-1052
五南文化廣場 40043 臺中市中區綠川東街 32 號 3 樓 (04)2221-0237
國家書店 10485 臺北市松江路 209 號 1 樓 (02)2518-0207

惜墨如玉：新品種寶石．臺灣墨玉
/ 方建能，林淑雅作．-- 初版．
-- 臺北市：臺灣博物館，民 103.03
面；29.7x21 公分
ISBN 978-986-04-0854-6(平裝)
1. 寶石 2. 玉器 3. 臺灣

357.89　　　　103005843